中国农业科技典籍翻译

ZHONGGUO NONGYE KEJI DIANJI FANYI (HANYING DUIZHAO)

（汉英对照）

编著

刘瑾

张雅琳

李鹏

陈国良

重庆大学出版社

图书在版编目（CIP）数据

中国农业科技典籍翻译：汉英对照 / 刘瑾等编著．重庆：重庆大学出版社，2025. 8. -- ISBN 978-7-5689-5021-3

Ⅰ. S；H315.9

中国国家版本馆 CIP 数据核字第 2025YP7257 号

中国农业科技典籍翻译（汉英对照）

编　著　刘　瑾　张雅琳　李　鹏　陈国良

责任编辑：牟　妮　　版式设计：牟　妮

责任校对：谢　芳　　责任印制：赵　晟

*

重庆大学出版社出版发行

社址:重庆市沙坪坝区大学城西路 21 号

邮编:401331

电话:(023) 88617190　88617185(中小学)

传真:(023) 88617186　88617166

网址:http://www.cqup.com.cn

邮箱:fxk@cqup.com.cn (营销中心)

全国新华书店经销

重庆新生代彩印技术有限公司印刷

*

开本:720mm × 1020mm　1/16　印张:24.25　字数:372 千

2025 年 8 月第 1 版　　2025 年 8 月第 1 次印刷

ISBN 978-7-5689-5021-3　定价:63.00 元

前言

农业科技典籍是指中国古代记述农业生产技术、生态管理、植物分类、食品加工等内容的科技类文献，如《吕氏春秋》《氾胜之书》《齐民要术》《茶经》《救荒本草》《本草纲目》《天工开物》和《农政全书》等。这些典籍既是中国古代农学知识的系统积累，也是中华文明生态智慧的重要体现。

本书以“农业科技典籍翻译”为核心，聚焦中国古代农学知识的典籍文本及其英译实践，旨在帮助学习者理解农业科技文献的语言风格、术语特点与文化内涵，并通过双语阅读与翻译训练，提升专业翻译能力与跨文化表达意识。本书内容共分为两个部分、九个章节，涵盖理论概述与实践方法，循序渐进，层层深入。

第一部分（第一章至第五章）侧重于农业科技典籍及其翻译的基本认知与理论支撑。第一章介绍科技典籍翻译的概念、原则、意义及译者素养要求；第二章聚焦农业科技典籍的文本特征、翻译历史与基本策略；第三章从翻译目的论、生态翻译观等角度探讨农业典籍中饮食、酿造等内容的翻译思路；第四章重点分析植物类典籍的术语翻译及赏析方法；第五章则围绕农田技术的翻译与典籍案例展开具体讲解。

第二部分（第六章至第九章）集中于翻译技巧的系统训练，包括词汇、句子、语篇和修辞等层面的翻译策略。第六章讲解术语、文化负载词和数量词的处理方法；第七章关注语序调整、增补、省略等句法转换技巧；第八章则分析照应、重复、连接等语篇衔接手段；第九章探讨比喻、拟人、押韵、排比等修辞的翻译方法，帮助学生掌握更具表现力的表达手段。

本书主要有以下几个特点：第一，每章配有针对性练习，包括术语辨析、翻译技巧分析、语法结构训练、段落翻译练习与简答题，以促进学生在掌握基础知识的同时，提升思维能力与实际操作能力。第二，注重理论引导结合实操训练，第一部分提供必要的翻译理论知识，第二部分则侧重翻译过程的分解训练。第三，注重跨文化理解与表达，在呈现农业技术细节的同时，强调文化差异与语境转换，引导学生从中外文化的对比视角出发，提升跨文化交际与表达能力。本书适合用于农业语言翻译、科技典籍阅读与文化传播类课程的本科或研究生阶段的教学使用，可作为课堂教学与自学参考。

本书编写过程中，选取了若干具有代表性的中国农业科技典籍译作，诚蒙众多译者前贤的心血之作及相关研究成果的启发与支撑，方能使本书得以成稿。在此，谨向所有原译作者与学术资料提供者表示诚挚谢意。由于编者水平所限，书中疏漏与不当之处在所难免，敬请专家学者及读者批评指正，以期在今后的修订中不断完善。

刘　瑾

2025年5月

目　录

第一部分　科技典籍及其翻译的介绍

第二部分　农业科技典籍翻译的方法

第一部分

科技典籍及其翻译的介绍

第一章　导　论

◎本章学习目标◎

1. 掌握科技典籍基本概念和分类
2. 掌握科技典籍的历史背景和发展过程
3. 了解科技典籍的翻译总原则

汪榕培、王宏将典籍定义为“中国清代末年1911年以前的重要文献和书籍”[①]。“我国应用翻译源远流长，特别在科技翻译方面，早在公元七世纪，就出现了天文学方面的译作。隋唐年间的《婆罗门天文经》《宿曜经》《都利聿斯经》等，是分别从梵文、康居文、波斯文翻译过来的。我国古时科学技术相对发达，科技译文在古籍中只是凤毛麟角。其中有的是随当时兴盛的佛经翻译同时传译过来的。中国科技典籍作为中华典籍与世界典籍的重要组成部分，不仅是中国科技文明与发展的载体，也为中国同世界的科技文化交流做出了重要贡献。在积极推动中国文化‘走出去’的时代背景下，科技典籍英译研究对于中国与世界均有着重要意义。中华典籍卷帙浩繁，分类方式繁多，常将其分为科技典籍和文化典籍”[②]（罗选民，李婕，2020）。科技典籍翻译是中国优秀科技文化走向国际舞台、建立中国科技话语体系和增强

① 汪榕培，王宏．中国典籍翻译[M]．上海：上海外语教育出版社，2009.

② 罗选民，李婕．典籍翻译的内涵研究[J]．外语教学，2020，41(6)：83-88.

中国文化软实力的重要途径。通过将科技典籍翻译成外语,我们可以向世界展示中国古代科学技术的辉煌成就,彰显中国作为科技强国的实力和创新能力。这种翻译不仅是简单的语言转换,更是一种文化传承和交流的方式。通过有效的翻译工作,我们可以将中国古代科技的智慧和经验传播到世界各地,促进不同文化之间的相互理解和合作。这种科技典籍翻译的努力对于推动中国科技文化的"走出去",构建中国科技话语体系以及提升中国文化的软实力具有重要意义。中国古代科技典籍数量浩如烟海,医学类达上万种,数学类有上千种,农学类四百余种[①](屈宝坤,2009)。

第一节　科技典籍翻译概论

中国科技典籍外译是传播中国古代科学技术、彰显国家力量和塑造民族形象的有效途径。在诸多体裁的文本翻译活动中,中国科技典籍外译涉及的翻译方法和技巧逐渐成为学者们频繁讨论的话题。其独特的文字表达、特殊的体裁规约及内在的文化属性很大程度上扩展了该领域翻译方法的种类和翻译技巧的使用。

科技典籍的译者需要具备深厚的语言功底和文化素养,以便能够恰当地选择合适的词汇和句式,使翻译结果既忠实于原文又符合目标语言的语言习惯。科技典籍往往具有严谨的逻辑结构和严密的论证方式,其内容涉及科学理论、技术实践、实验方法等方面。翻译者在进行外译时需要准确把握原文的逻辑关系和论证过程,以保证翻译结果的准确性和连贯性。同时,他们还需要根据目标读者的背景和需求,适当调整语言风格和表达方式,使翻译结果更易于理解和接受。此外,科技典籍不仅是科学技术的传承和积

① 屈宝坤.中国古代著名科学典籍[M].北京:中国国际广播出版社,2009.

累，更是中国古代文化的重要组成部分。其中融入了中国传统哲学、道德观念、审美情趣等方面的思想内容。翻译者在进行外译时需要对这些文化内涵有深入的理解，以便能够准确传达原文所包含的思想和价值观。同时，他们还需要注意避免将自己的文化观念和价值观强加于译文中，以保持原文的原汁原味。

第二节 科技典籍翻译的概念

“科技典籍”虽未有统一明确的定义，但已有学者在整理、借鉴和融合相关联概念的基础上力图对其加以定义[①]（刘性峰，王宏，2016）。许明武和王烟朦两位学者对科技典籍的时间划分、书写主体、内容与载体形式等均有较为明晰的界定，即中国科技典籍为“1840年以前由域内人士直接撰写的以科学技术为主要内容的重要文献，载体形式多样，涵盖数学、天文、物理、化学、地学、生物学、农学、医学、技术和综合十大类”[②]（许明武，王烟朦，2017）。

科技文体“具有信息明确、逻辑连贯、陈述客观等文体特征”[③]。“翻译科技典籍时应尽可能地遵循上述体裁规约和期待规范，‘述’的作用也是服务于此。‘述’是在原文的信息不明之处、逻辑不清之处、科学性存疑之处和译文的起承转合之处，添加外文书写，增强译文的明晰性、逻辑性、客观性和连贯性”[④]。“翻译科技典籍时应当依托权威释本和最新相关‘研而著’成果，同时参照其他释本乃至善本。这既是准确地理解原文的应然要求，也是传播

① 刘性峰，王宏．中国古典科技翻译研究框架构建[J]. 上海翻译，2016(4)：72，77-81，94.

② 许明武，王烟朦．中国科技典籍英译研究（1997—2016）：成绩、问题与建议[J]. 中国外语，2017，14(2)：96-103.

③ 梅阳春．古代科技典籍英译——文本、文体与翻译方法的选择[J]. 上海翻译，2014(3)：70-74.

④ 梅阳春．西方读者期待视阈下的中国科技典籍翻译文本构建策略[J]. 西安外国语大学学报，2018(3)：102-106.

现实价值的必要前提。中国科技典籍并非按照西方的科学知识分类和科学逻辑来编写,而且其中包含不少文学性、哲学性甚至神话迷信元素”[①]。

科技典籍翻译是指将科技领域的典籍、经典著作等原文翻译成其他语言的过程。科技典籍翻译的目的是促进科技领域的交流和发展,让更多的人了解和掌握科技领域的知识和技术。科技典籍翻译可以帮助人们跨越语言和文化的障碍,促进不同国家和地区之间的合作和交流。

第三节　科技典籍翻译的原则

马清海曾提出科技翻译的原则有三点:一、准确:理解和表达科技内容、科技概念(特别是科技术语)、语言形式、逻辑关系、符号公式、图标数字等都要准确无误,要忠实于原文。二、简洁:遣词用句、行文简洁明了,精练通顺。三、规范:语言、文字、术语、简称、符号、公式、语体、文章体例、计量单位等都要规范统一,符合国家和国际标准。[②]除此以外,严复在《天演论》提出了翻译的基本标准和原则:“信、达、雅。”

本教材在此基础上拓展为四个方面:一是翻译的准确性;二是翻译的通顺流畅;三是规范性;四是文化的适应度。

总之,在科技翻译中,准确性是至关重要的。科技文献中包含大量专业术语和概念,译者需要准确理解原文内容,并确保翻译准确无误。任何错误或误解都可能导致信息传递的失真,影响读者对文本的理解和应用。因此,科技译者需要深入研究原文,确保准确理解每一个专业术语和概念,并将其精准地转化为目标语言,以保证翻译的准确性。

① 张保国,周鹤.石声汉的农学典籍译介模式及其启示[J].解放军外国语学院学报,2022,45(5):119-127.

② 马清海.试论科技翻译的标准和科技术语的翻译原则[J].中国翻译,1997(1):2.

除了准确性，科技翻译还需要保持通顺流畅。科技文献往往结构复杂，逻辑严谨，译者需要确保翻译文本在语言表达上通顺流畅，符合目标语言的语言习惯和表达习惯。流畅的翻译能够提升读者的阅读体验，使信息传递更加清晰和有效，帮助读者更好地理解文本内容。

科技翻译需要符合相关的规范和标准。译者应遵循专业的翻译准则和规范，确保翻译文本符合语言规范和专业要求。规范性的翻译能够提高文本的可读性和可信度，让读者更容易理解和接受翻译内容。同时，规范性的翻译也有助于保持翻译领域的权威性。

著名科技典籍翻译家刘仙洲教授编订了《英汉对照机械工程名词》（商务，1947），为翻译及统一中国机械工程名词做出了贡献。他广泛收集明朝以来有关的出版物30多种，调查当时机械工程界已有的各种名称，工人是怎么叫的，日文又怎么译法，然后逐个做成卡片，按照“从宜”“从熟”“从简”“从俗”的四大原则，从中选定一个。①

清末语言学家、翻译评论家马建忠（1845—1900）熟悉西方政治、学术与文化。1894—1895年中日战争爆发后，他目睹国难时艰，认为我国欲知彼知己，必须发展翻译事业。光绪二十年（1894年），他向清廷呈奏折《拟设翻译书院议》。在百年前的奏折中提出了“善译”的翻译标准。他要求“译成之文，适如其所译而止，而曾无毫发出入于其间”。为达到此标准，译者应做到“一书到手，经营反复，确知其意旨之所在，而又摹写其神情，仿佛其语气，然后心悟神解，挥笔而书”。他实质上提出了译文应与原作形似、意似、神似。他强调译者平日对原文语言与译文语言必须有很深的修养，并对语法学、词源学、文体学、修辞学、语音学等学科也都应有较深的研究。②

---

① 林煌天，贺崇寅．中国科技翻译家辞典[K].上海：上海翻译出版公司，1991.

② 许明武，王烟朦．中国科技典籍英译研究（1997—2016）：成绩、问题与建议[J].中国外语，2017，14(2)：96-103.

现代马克思主义理论家、教授、马克思主义著作翻译家李达(1890—1966)在1954年某次座谈上针对时弊发表了题为《谈谈翻译》的讲话,指出译者必须提高中文修养与政治水平,翻译自己所熟悉的东西,采取对人民负责的态度。他特别强调学习马列主义和毛泽东思想在翻译(尤其是社会科学翻译)中的重大作用。这番语重心长的谈话对我国今后的翻译工作者仍然具有指导意义。①

"康有为翻译思想缘于国势之日孱,发乎忧世忧民之襟怀,期以效泰西法东洋而达于华夏之强盛同国性之勃兴。以译启民智,以译觉乱世,以译救亡国,以译臻大同,康子翻译思想之文化民族主义与文化世界主义意蕴昭昭明甚矣。康子'以译启民、以译觉世、以译救国、以译大同'翻译思想之价值旨趣亦民族,亦世界,散发着浓厚之文化民族主义与文化世界主义气息。文化民族主义一为经,文化世界主义一为纬,交错贯穿于康子翻译思想形成之始末,且互为滋养,互为辉映。康子倡译日书、西书以挽民族于狂澜,其文化民族主义之情怀可鉴;至若欲挽民族于狂澜,以积大同蓝图之跬步,其文化世界主义之情怀可表。作为观念性之存在,康子翻译思想发乎于文化民族主义而止于文化世界主义。换言之,文化民族主义启康子翻译思想之端,而文化世界主义继康子翻译思想之绪,乃文化民族主义意蕴之升华,亦为康子翻译思想之最终归宿。"②因此,科技翻译需要考虑文化的适应度。不同的语言和文化背景会影响翻译文本的表达方式和理解方式,译者需要根据目标读者的文化背景和习惯进行调整,确保翻译内容在文化上的适应度。适应目标文化的翻译能够增强读者的接受度和理解度,使翻译内容更具有针对性和影响力。

---

① 许明武,王烟朦.中国科技典籍英译研究(1997—2016):成绩、问题与建议[J].中国外语,2017,14(2):96-103.

② 王惠琼,孔令翠.20世纪前海外中国农业科技典籍译介研究[J].外国语文,2022,38(3):108-115.

第四节 科技典籍翻译的意义

科技典籍翻译的意义在于推动科技知识的传播、促进科技进步和推动全球科技合作。科技典籍翻译及其有效的对外传播是全面推动中国文化对外传播的途径之一,对构建我国对外翻译话语、维护文化生态平衡、提升国家文化软实力、增强民族自信心将起到重要作用[①](王燕,李正栓,2020)。科技典籍翻译的意义主要有以下几个方面:

第一,有助于科技知识的对外传播。让更多的科技研究人员和科技爱好者了解到最新的科技发展和研究成果,推动中国科技知识与技能的全球化传播。科技典籍翻译融合了不同国家和地区的科技成果和知识,为科技研究人员提供了更广阔的学习和借鉴资源,促进科技进步和推动全球科技合作。

第二,有助于文化的传承,塑造中国国际形象。中国科技典籍是一个国家或地区的文化遗产,通过翻译可以将这些典籍传播到世界各地,保护和传承它们的文化价值,传播中国科技典籍中的智慧和经验,提高中国软实力。

第三,有助于国际文化交流。通过翻译,不同文化背景的人们可以了解和理解其他文化中的思想、价值观和传统。这有助于促进不同文化之间的交流和对话,增进彼此的理解和尊重,减少文化冲突和误解。

具体到农业科技典籍翻译的意义,本书参照[②](王惠琼,孔令翠,2022)提出的6个方面的意义如下:

---

① 王燕,李正栓.《大中华文库》科技典籍英译与中国文化对外传播[J].上海翻译,2020(5):53-57,94.

② 王惠琼,孔令翠.20世纪前海外中国农业科技典籍译介研究[J].外国语文,2022,38(3):108-115.

(1)推动了译介国的农业技术应用与科技革命

《天工开物》中的制糖技术与《农政全书》中的农业技术推动了整个日本农业技术的发展和农业生产力的提高①(潘吉星,1990),在日本实现了农业现代化后仍然受到重视②(杨直民,1985),为回归自然农法及推进现代有机农业、环境保全型农业的发展提供了重要借鉴,做到了古为今用③(胡火金,2016)。

(2)促进了译介国桑蚕业发展

促进了世界经济的发展和贸易的繁荣④(张萌 等,2015)。日本丝织业长期学习与借鉴中国先进的丝纺织技术,推动了蚕丝业蓬勃发展,被称为日本经济起飞的功勋产业。杜赫德和儒莲等的译介不但促进了法国蚕丝业发展,还有助于其成为欧洲强国⑤(潘吉星,2013)。

(3)影响了译介国农业治理政策的制定与理念的更新

日本本草学大师松冈玄达参考《农政全书·荒政》写作的《救荒本草》及《野菜谱》对日本制定备荒与救荒政策发挥了重要作用。法国重农学派学说的产生与《农政全书》等的译介有着深厚的中国渊源⑥(谈敏,2010)。法国重农学派创始人魁奈非常赞赏中国的重农主义和历朝历代重视农业的政策,提倡以农为本,认为只有农业能够增加财富,因而要求改变当时普遍存在的轻视农业发展的思想⑦(朱高正,2001)。

(4)激发了东亚译介国崇尚实学的经世致用思想的形成与发展

日本实学派学者佐藤信渊依据《天工开物》提出富国济民的技术哲学和

---

① 潘吉星.宋应星评传[M].南京:南京大学出版社,1990.

② 杨直民.农业科学技术史研究的蓬勃发展[J].中国农史,1985(3):55-63.

③ 胡火金.日本学者对中国农业史的研究[J].史学月刊,2016(11):124-129.

④ 张萌,刘俊仙.丝绸之路与古代中国蚕桑技术的外传[J].中国民族博览,2015(8):184-186.

⑤ 潘吉星.《天工开物》在国外的传播和影响[N].北京日报,2013-01-29.

⑥ 谈敏.法国重农学派学说的中国渊源[M].上海:上海人民出版社,2010.

⑦ 朱高正.中国文化对西方的影响[J].自然辩证法研究,2001(8):49-55.

技术经济学思想“开物之学”[①](都贺庭钟,1771)。

(5)促成了译介国茶文化与茶产业的兴起

在欧洲,1662年,酷爱饮茶的葡萄牙凯瑟琳公主嫁给英国查尔斯二世后在英国朝廷和王公贵族间刮起饮茶之风。18世纪中后期,茶开始进入了欧洲平民的生活[②](王思明 等,2017)。到19世纪,英国人开始尝试在红茶中加入玫瑰、薄荷、柠檬等,有时还会加些鲜奶和糖(袁梦瑶 等,2019),从而开创了新的茶文化。总之,对茶文化的译介不但为世界贡献了健康的饮食习惯、良好的养生方式、美好的精神享受、恭敬的待客之道,还提供了发展经济之道(施由明,2018),推动了茶叶生产与消费。

(6)提升了中国古代农业科技的国际地位和学术影响

提升了中国古代农业科技在世界上的形象,促进了中国农学的东渐西传,还丰富了汉学的内容。

随着我国综合国力和国际影响力大幅提升,对外传播中华文化,让世界了解中华文化、读懂中国的需求显著增强。典籍外译不仅契合国家鼓励中华优秀传统文化“走出去”与“一带一路”倡议的时代精神,还是向世界传播中国理论、中国思想、中国声音的有效路径。[③](孙美娟,2022)

第五节 科技典籍译者的素养要求

科技典籍需要翻译人员具备扎实的专业知识和优秀的语言功底。这项工作要求翻译人员不仅要理解原文的科技内容,还要准确传达其中的专业术语和信息,确保翻译准确无误、通顺流畅。翻译人员还需考虑到不同文化

---

① 都贺庭钟.天工开物菅生堂本序.菅本书首[M].大阪:菅生堂.1771.

② 王思明,李昕升.农业文明:丝绸之路上“行走”的种子[N].中国社会科学报,2017-03-02.

③ 孙美娟.中国社会科学网-中国社会科学报[N].2022-06-22.

背景下的差异,避免出现文化冲突或误解。在翻译过程中,要注重保持原文的科技性和专业性,同时适当考虑目标读者的理解程度,确保翻译结果既准确又易于理解。

(一)扎实的双语翻译转换能力

一名优秀的科技典籍翻译人员需要具备扎实的语言基础和翻译技能,能够准确、流畅地将信息从一种语言转换为另一种语言。这种能力不仅需要熟练掌握源语言和目标语言的语法、词汇和表达方式,还需要具备深入理解不同语言文化背景的能力。双语翻译转换能力的核心在于准确传达原文的含义,同时保持翻译结果的自然流畅。翻译人员应该注重细节,确保术语和表达在两种语言中的准确对应,避免信息失真或歧义。此外,科技典籍翻译人员还需要具备逻辑思维和条理性,能够清晰地组织和呈现翻译内容,使读者能够准确理解原文信息。

(二)"外语+X"复合式专业能力

科技典籍翻译工作对译者的要求不仅在于其外语水平,还需要具备"外语+X"复合式专业能力。译者需要具备扎实的外语基础,能够准确理解和表达源语言的含义。这包括语法、词汇、语言表达能力等方面的要求。除了外语能力,译者还需要深入了解某一专业领域的知识,如科技、工程、医学等领域。科技典籍翻译中,译者需要具备跨文化沟通能力,能够理解不同文化背景下的差异,避免文化冲突或误解,通过学习跨文化交流理论、参与国际交流项目、了解不同文化的习俗和价值观等方式来提高跨文化沟通能力。

在农业科技典籍翻译过程中,术语的翻译是最重要的一环,如果不仔细查询术语来源就会出现一些术语偏差错误。如 Crop rotation 被译成"作物轮种",实际上应该是"轮作",因为"轮作"更准确地描述了一种农业实践,即在同一块土地上轮换种植不同作物。Cover crops 被译成"覆盖作物",实际上应该是"绿肥作物"或"覆盖植被",因为"绿肥作物"更准确地描述了这类作物的功能和种植方式。Food security 被译成"食品安全",实际上应该是"粮食安全"或"食品保障",因为"粮食安全"更具体地指代保障人们获得足够的营养食物。因此,在翻译前应该通过查阅多方权威发布确定术语的概念。

(三)翻译职业素养

科技典籍兼具科学性与文学性,因此对译者的要求更高,除了需要具备扎实的科技领域知识,包括相关学科的专业术语和概念,以便准确理解原文内容并准确翻译,更需要具备严谨的工作态度和细致的工作习惯,对原文内容进行逐字逐句的分析和翻译,确保翻译准确无误。尤其对于科学领域术语翻译准确程度的把握,科技典籍译者需要具备清晰的逻辑思维能力,能够准确理解原文的逻辑结构和论证思路,并在翻译过程中保持逻辑连贯。科技典籍译者需要具备良好的沟通能力,能够与专业领域的专家进行有效的沟通和交流,以便更好地理解原文内容并进行准确翻译。为了成为优秀的科技典籍译者,译者需要不断努力提升自己的专业知识水平,保持严谨和细致的工作态度,加强逻辑思维能力的训练,并提升沟通能力和与专业领域专家的交流能力。通过不断努力和提升,更好地胜任科技典籍翻译工作,提高翻译水平和专业素养。

(四)翻译技术运用能力

科技译者在应对科技发展与社会变化带来的挑战时,需要保持学习热情,培养专业敏感度和综合素质。他们应具备良好的网络搜索能力,准确鉴别资料价值,确保翻译质量。新兴翻译技术的发展提升了工作效率,使译者能更专注于创新,提高成果质量。科技译者需调整职业规划,充实知识,增强核心竞争力,成为复合型人才。通过持续提升专业素养,科技译者将开拓更广阔的职业视野,成就更丰富的职业生涯。在应对行业挑战时,科技译者的翻译技术运用能力至关重要。

在科技翻译领域,特别是在农业领域的英语翻译中,搜索技能的重要性更加突出。例如,在农业科技文献情报工作中,译者需要通过搜索技能查找农业领域的最新研究成果、种植技术和市场趋势,以支持农业文献的翻译工作。译者需要善于筛选各种来源,包括学术期刊、专业报告和农业数据库,以获取有用的资料并准确理解其中的专业术语和概念。另外,译者还需要能够鉴别资料的可靠性和权威性,确保翻译内容的准确性和专业性。通过搜索技能的提升,农业英语翻译者可以更好地应对农业领域复杂多变的信息,并在翻译过程中准确传达农业科技知识,为农业领域的发展和交流做出贡献。

第六节　翻译技术辅助下的科技典籍翻译

科技典籍是人类智慧的结晶,是科技发展的重要基石。然而,由于语言的壁垒,许多优秀的科技典籍的国际传播受到限制。翻译技术的出现为科技典籍的跨语言传播提供了有力的支持。机器翻译是指通过计算机程序将一种语言的文本自动翻译成另一种语言的技术。在科技典籍翻译中,机器翻译可以大大提高翻译效率和准确性。首先,机器翻译可以快速处理大量的科技典籍,节省了人力和时间成本。其次,机器翻译可以减少翻译过程中的人为错误,提高翻译质量。此外,机器翻译还可以实现实时翻译,使科技典籍的内容能够及时传播和应用。因此,人机结合成为科技典籍的翻译新途径,计算机辅助翻译(Computer Assisted Translation,CAT)是指通过计算机软件辅助翻译人员进行翻译的技术。在科技典籍翻译中,CAT可以提高翻译的准确性和一致性。首先,CAT可以建立术语库和翻译记忆库,帮助翻译人员在翻译过程中快速查找和应用专业术语和翻译记忆。其次,CAT可以提供自动校对和术语一致性检查等功能,减少翻译过程中的错误和不一致。此外,CAT还可以支持多人协作翻译,提高翻译效率。然而,CAT在科技典籍翻译中也存在一些挑战。CAT的效果依赖于术语库和翻译记忆库的质量和完整性。如果科技典籍的领域特殊或者是新兴领域,可能缺乏相应的术语和翻译记忆,导致CAT的效果不理想。语料库的建设与运用对科技典籍翻译的辅助作用不容小觑,常见的典籍语料库和术语库如下。

名称	简介	创办者
中华文化国际传播网	网站以“汇聚文化资源,推动文明互鉴”为宗旨,坚持以文载道、以文传声、以文化人,向世界阐释推介具有中国特色、体现中国精神、蕴藏中国智慧的优秀文化,为全球中华文化的传承者、守护者和爱好者提供交流互鉴的平台。	中国外文局文化传播中心
中国哲学书电子化计划	中国哲学书电子化计划是一个线上开放电子图书馆,为中外学者提供中国历代传世文献,力图超越印刷媒体限制,通过电子科技探索新方式与古代文献进行沟通。收藏的文本已超过三万部著作,并有五十亿字之多,故为历代中文文献资料库最大者。	英国杜伦大学 Donald Sturgeon

续表

名称	简介	创办者
中国农业历史与文化(农业典籍)	该平台包含所有农业相关的农业技术,农业历史与文化,农业新闻,农器图谱、农业典籍40余部著作等。	中国科学院自然科学史研究所
汉典古籍	汉典古籍为免费古籍文库,集而公诸同好,并献天下,以承古风,至今得文三万八千五百二十九。	未知
汉语世界	《汉语世界》(TWOC)的主旨是向世界传播汉语和中国文化,有英文纸媒和电子杂志。作为当代中国最权威的研究资源之一,它以其对中国社会的深度报道、客观性和以人为本的态度而闻名,包括人物、当代潮流、历史、风俗、文化交流等。	商务印书馆
术语在线	术语在线提供各学科领域的专业术语、新词、资讯等信息,支持术语接口、标注、空间等功能。	全国科学技术名词审定委员会
中华思想文化术语库	主要提供中国传统思想文化典籍的经典外译,均为中外对照。	外语教学与研究出版社
中国关键词	“中国关键词”项目是以多语种、多媒体方式向国际社会解读、阐释当代中国发展理念、发展道路、内外政策、思想文化核心话语的窗口和平台,是构建融通中外的政治话语体系的有益举措和创新性实践。	中国外文局和中国翻译研究院发起,中国翻译协会和中国外文局对外传播研究中心具体组织实施。

续表

名称	简介	创办者
Food and Agriculture Organization of the United Nations	联合国粮食及农业组织(FAO)是联合国的一个专门机构,负责领导全球消除饥饿的国际努力。其目标是实现全民粮食安全,确保人们能够经常获得足够的高质量食物,从而过上积极、健康的生活。FAO拥有195个成员(194个国家和欧盟),并在全球130多个国家开展工作。	联合国粮农组织
识典古籍	识典古籍是古籍数字化三十年以来,中国大陆目前面向大众开放的最大规模的古典文献阅读与整理平台,在AI数字环境下推广中国传统文化。	北京大学数字人文研究中心与抖音集团联合创造
Chinese History Digest	该英文网站主要分为两部分:第一部分按时间顺序讲解中国各个历史时期;第二部分详细介绍在中国历史上具有重要意义的古迹景观。	Prof. Kenneth J. Hammond (New Mexico State University)
中国文化网	该网站是我国对外和对港澳台文化工作领域内的权威资讯平台,中文版设置了头条、动态、专题、文贸等栏目,整合了海外中国文化中心、"欢乐春节"、中国对外文化交流协会、中华文化联谊会等专栏,是及时了解中外文化新闻事件、发展动态、交流情况、优势资源的窗口。	中华人民共和国文化和旅游部国际交流与合作局(港澳台办公室)

续表

名称	简介	创办者
中国古典文献资源导航系统	网站坚持“学术乃天下公器”之精神,面向文献学、古代文学、古代史、古文字学等传统文史领域的研究者、初学者(尤其是硕博研究生)和广大民众爱好者提供免费、公益性服务。	唐宸(安徽大学)
重要概念范畴表述外译发布平台	该平台及时发布习近平总书记和中央文件提出的新概念新范畴新表述,以及当下中国的新词、热词、敏感词、特色词等的多语种外文表达,以权威专业、及时有效为特征,服务于对外传播翻译工作实践和中国特色话语对外翻译标准化建设。	中国外文局、当代中国与世界研究院、中国翻译研究院
中国特色话语对外翻译标准化术语库	该术语库为中国思想文化、中国特色话语的对外翻译、传播,提供统一、规范的多语种术语及相关知识的数据资源查询服务,有效服务对外传播翻译工作。	中国外文局、中国翻译研究院

借助于以上语料库和术语库,科技典籍翻译提高了效率与准确度。双语语料库有利于典籍翻译爱好者的学习与运用。在科技典籍翻译中,人工智能(Artificial Intelligence,AI)可以应用于语义理解、语言生成和机器学习等方面,提高翻译的质量和效率,可以通过深度学习等技术提高机器翻译的准确性和流畅度,使翻译结果更加自然和准确,同时实现语义理解和上下文分析,帮助机器翻译系统更好地理解科技典籍的内容和意义。此外,人工智能还可以实现自动摘要和文本生成等功能,为科技典籍的翻译和理解提供

更多的支持。然而,机器翻译在科技典籍翻译中也存在一些挑战。首先,科技典籍的内容通常较为复杂和专业化,需要具备丰富的专业知识背景和语料库。机器翻译系统在处理专业术语和复杂句式时可能存在困难,导致翻译结果不准确。其次,科技典籍的翻译需要考虑上下文和语境,而机器翻译系统往往只能进行局部翻译,无法全面考虑文本的整体意义。如今在各种CAT工具的辅助下,科技典籍翻译借助于已有的专业语料库及术语库,实现了译前—译中—译后的工作化流程。翻译技术可以提供术语管理工具,帮助译者收集、整理和管理科技术语,确保术语在不同文档中的一致性和准确性。译前通过语帆术语宝、SDL MultiTerm Extract、SynchroTerm等工具进行术语提取,保证翻译的一致性。通过CAT工具翻译也便于译后对同类主题的译文进行参考,提高效率。因此,科技典籍译者在人工智能时代有必要掌握Trados、Déjà Vu与MemoQ等CAT软件。

练 习

1.思考题。

1)科技典籍翻译在促进中国科技进步方面扮演着重要角色,请谈谈你对这一观点的看法。

2)科技典籍翻译中的准确性和可读性有时会产生冲突,请讨论如何在翻译过程中平衡这两个方面。

3)科技典籍译者需要具备哪些科技知识和语言能力?请列举并解释其重要性。

4)在翻译技术不断发展的背景下,科技典籍翻译面临着哪些挑战?你认为应如何应对这些挑战?

5)对于科技典籍的翻译工作,你认为译者应该具备怎样的思政素养和批判性思维?为什么这些素养和思维对于科技典籍翻译至关重要?

2. 句子翻译。注意英汉语言的转换技巧,以及句子中客观事实的描述。

1)我国不但有悠久的农业历史,而且在长期的农耕实践中产生了丰富的农学思想和农学典籍。

2)奠基于春秋战国时期的中国古代农学思想,以整体、辩证、发展为特点,强调天地人之间的和谐,成就了中华农业的长盛不衰。

3)古代农业生态理念的建立,对于协调发展与环境之间、资源利用与保护之间的矛盾,形成社会经济发展的良性循环具有重要的历史意义和现实意义。

4)古代农业典籍是我国先民留给后人的一项重要财富,它们的搜集、编纂和校注是中国农业遗产研究的基础,前期的保护工作大多致力于此。

5)农业遗产与农耕文化互为表里,是先民万年农耕实践的智慧结晶,是祖先留给我们后人的宝贵财富,具有跨越时空的永恒价值。

6)迁西板栗形端粒匀,色泽鲜艳,香糯甘甜,营养丰富,品质居世界板栗之首。

7)释超全所著《武夷茶歌注》中有这样的说法:“建州有一位老人最早开始进献山上采来的茶叶。据说老人去世后变成了山神,喊山茶的习俗由此而来。”

8) “People could easily buy iced beverages on the street during the Song Dynasty (960–1279).”

9) “Ice water sold during the Song Dynasty was often added to mung beans or licorice to help prevent heat stroke.”

10) “The climate in the Yellow River area in the Tang Dynasty was warmer and moister, very different from today.”

11) “When rich noble people had cherries with cheese and cane syrup, they often used plates and bowls made of gold or colored glaze to make the fruit look more mouthwatering.”

12) John Knoblock and Jeffrey Riegel have provided us with a marvelous work in this first rate annotated translation of the complete extant text, as well as extensive notes and appendices... In the past, we have found ourselves in debt to the late John Knoblock for his *Xunzi*. Now we are further indebted to him and Jefrey Riegel for this splendid scholarly work.

—*American Jounal of Chinese Studies*

3.段落翻译。修改完善以下译文,让译文更易于目的语读者阅读。

《吕氏春秋》是在秦国丞相吕不韦主持下,集合门客们编撰的一部黄老道家名著。成书于秦始皇统一中国前夕。此书以儒家学说为主干,以道家理论为基础,以名、法、墨、农、兵、阴阳家思想学说为素材,熔诸子百家学说为一炉,闪烁着博大精深的智慧之光。吕不韦想以此作为大一统后的意识形态。但后来执政的秦始皇却选择了法家思想,使包括道家在内的诸子百家全部受挫。《吕氏春秋》集先秦道家之大成,是秦道家的代表作,全书共分二十六卷,一百六十篇,二十余万字。《吕氏春秋》分为十二纪、八览、六论,注重博采众家学说,是以道家思想为主体兼采阴阳、儒墨、名法、兵农诸家学说而贯通完成的一部著作。其主要的宗旨属于道家。所以,《汉书·艺文志》等将其列入杂家。高诱说《吕氏春秋》"此书所尚,以道德为标的,以无为为纲纪",这说明最早的注释者早已点明《吕氏春秋》以道家为主导思想之特征。

The Spring and Autumn Annals of Lü Buwei is a masterpiece of Huang Lao-dao, which was compiled by a group of scholars under the auspices of Lü Buwei , the prime minister of Qin State. The book was written on the eve of Qin Shihuang's unification of China. This book is based on Confucianism, Taoism, the thoughts and theories of Ming, Fa, Mo, Nong, Bing and Yin and Yang, melting hundred schools of thought's theory into a furnace, shining with profound wisdom. Lü Buwei wanted to take this as the ideology after the unification. However, Qin

Shihuang, who was in power later, chose legalism, which frustrated all hundred schools of thought, including Taoism. *The Spring and Autumn Annals of Lü Buwei* is a masterpiece of Taoism in the pre-Qin period. It is divided into 26 volumes, 160 articles and more than 200,000 words.It is a work structured into twelve Ji (Records), eight Lan (Examinations), and six Lun (Discussions). It emphasizes the broad collection of various schools of thought, with Daoism as its core while integrating and synthesizing ideas from Yin-Yang, Confucianism, Mohism, Logicians, Legalism, Military, and Agricultural schools. But the main purpose belongs to Taoism. Therefore, Hanshu Yiwenzhi listed it as a sage. Gao You said that the book *The Spring and Autumn Annals of Lü Buwei* is "moral-oriented and inaction-oriented", which shows that the earliest annotators have already pointed out that the book *The Spring and Autumn Annals of Lü Buwei* is dominated by Taoism.

延伸阅读

What Did Ancient Chinese People Eat in Summer?

On hot summer days, few people have a good appetite. Ice cream and fruit sound more refreshing than a bowl of hot noodles. In the past, ancient Chinese people also enjoyed "ice cream" and specially prepared fruit in summer.

Su shan

Su shan (酥山) is a dish that looks like crushed ice with milk and butter. Poet Wang Lingran from the Tang Dynasty (618–907) described how people made and enjoyed *su shan* in his poem *Ode to Su He Shan*. According to the text, sugar was added to *su shan* that was shaped into many forms. Sometimes, *su shan* was decorated with flowers and leaves to make it more beautiful. "It is neither solid, nor watery and disappears once it touches teeth", the poet wrote. The *su shan* in his poem was thought by many scholars to be an early form of ice cream. However, only the royals and nobles were lucky enough to have large iceboxes to create the delicacy in summer.

Ice and iced beverages

Although there were no fridges in ancient times, iceboxes to store ice cubes in summer had become very common in rich families by the Tang Dynasty. At that time, people either had ice water or crunched shaved ice. People could easily buy iced beverages on the street during the Song Dynasty (960–1279). As a community service, some rich people even provided free ice water and medicine on the street for free. Ice water sold during the Song Dynasty was often added to mung beans or licorice to help prevent heat stroke.

Cherries with cheese and cane syrup

During the Wei, Jin and Southern and Northern dynasties (220–581), dairy food appeared more frequently on ancient Chinese people's tables.

People made three kinds of cheese. One was called *tian lao*, or sweet cheese, which tasted like cheese yet looked like yogurt. Another is *gan lao*, or dried cheese, similar to solid cheese eaten today. The last one is *cu lao*, a kind of half-sour, half-sweet cheese, like yogurt.

The climate in the Yellow River area in the Tang Dynasty was warmer and moister, very different from today. Many cherry trees were planted there at that time. The fruit was common in early summer and people often added cheese and cane syrup to the cherries. Lu You, a noted poet from the Song Dynasty, once wrote: "eat cherry, peach and cheese at the same time." Thus in the Song Dynasty, when ancient Chinese people sent cherries as gifts, cheese was often presented together. When rich noble people had cherries with cheese and cane syrup, they often used plates and bowls made of gold or colored glaze to make the fruit look more mouthwatering.

1.根据以上内容,完成以下阅读理解。

1) What is "su shan" in ancient Chinese cuisine?

A. A type of ice cream made with milk and butter

B. A dish made with crushed ice and sugar

C. A type of fruit salad with flowers and leaves

D. A cold beverage made with ice water and mung beans

2) How did ancient Chinese people store ice cubes in summer?

A. In fridges

B. In iceboxes

C. In underground cellars

D. In special containers made of bamboo

3) During which dynasty did people start to sell iced beverages on the streets?

A. Tang Dynasty

B. Song Dynasty

C. Wei Dynasty

D. Jin Dynasty

4) What type of cheese was commonly eaten with cherries and cane syrup in ancient China?

A. Sweet cheese

B. Dried cheese

C. Half-sour, half-sweet cheese

D. All of the above

5) How did rich noble people in the Song Dynasty present cherries with cheese and cane syrup?

A. On simple wooden plates

B. On plates and bowls made of gold or colored glaze

C. In bamboo containers

D. In ceramic dishes

2. 词汇学习。

上农: the supreme importance of agriculture

任地: the requirements of the land

辩土: discriminating types of soil

审时: examining the season

开垦: reclamation: The process of developing and utilizing barren land, swamps, etc.

开发利用: development and utilization

水利: water conservancy

荒政: famine relief policies

救灾措施: disaster relief measures

abelmoschus moschatus 黄葵;山油麻

truck farm 蔬菜农场

truck farming 商品蔬菜栽培

truck growing (蔬菜)运销栽培

拓展知识

李善兰

李善兰(1811年1月22日—1882年12月9日),原名李心兰,字竟芳,号秋纫,别号壬叔,浙江海宁人,是清代卓越的数学家、天文学家、力学家、植物学家及翻译家。他自幼对数学表现出极高的兴趣,十岁时便主动阅读《九章算术》,并展现出过人的领悟能力。成年后,他深入研究数学理论,并在上海参与了西方科学著作的翻译工作,在八年时间内完成了八十余卷译作。1868年,他受聘于京师同文馆,担任天文算学馆总教习,直到去世。他的数学研究体系主要汇编于《则古昔斋算学》一书,其中《方圆阐幽》《弧矢启秘》《对数探源》等作品探讨了幂级数展开及其应用。

李善兰的数学贡献具有重要价值,他推导出二次平方根的幂级数展开式,并深入研究了三角函数、反三角函数和对数函数的展开公式,这些成果构成了19世纪中国数学领域的重要进展。他还提出了"尖锥术",利用尖锥的面积作为数学表达工具,并通过求和方法解决复杂的计算问题。尽管当时他尚未正式接触微积分,但其推导结果已具备定积分的基本特征,并成功将该方法应用于对数函数的幂级数展开,推动了中国数学的发展。

1852—1859年,李善兰在上海墨海书馆与英国传教士、汉学家伟烈亚力等人合作翻译出版了《几何原本》后九卷,以及《代数学》《代微积拾级》《谈天》《重学》《圆锥曲线说》《植物学》等西方近代科学著作,又译《奈端数理》(即牛顿《自然哲学的数学原理》)四册(未刊),这是解析几何、微积分、哥白尼日心说、牛顿力学、近代植物学传入中国的开端。

《植物学》是晚清时期传入中国的第一本系统介绍西方植物学知识的译著,在近代中国植物学史上占据着重要的地位。该书为李善兰与英国在华传教士韦廉臣(Alexander Williamson,1829—1890)合译,原著由英国植物学家林德利(John Lindley,1799—1865)所著的《植物学基础》(*Elements of Botany*)。该书的出版对于西方植物学在中国的传播和近代中国植物学科学体系的建立起到了基础性的作用。墨海书馆于1858年出版发行了这部八卷本的著作。《植物学》的问世填补了中国植物学知识的空白,为中国学界引入了当时最先进的西方植物学理论和研究成果。这部译著的出版不仅推动了中国植物学的发展,也为后来的翻译工作和学术交流打下了基础。

李善兰的翻译工作是有独创性的,他创译了许多科学名词,如"代数""函数""方程式""微分""积分""级数""植物""细胞"等,匠心独运,贴切恰当,不仅在中国流传,而且东渡日本,沿用至今。咸丰九年,《几何原本》后七卷一起刊行于世,因战争,原版被毁,后又于同治五年由李鸿章重刊,光绪十四年上海六合书局又石印出版。《重学》一书的翻译出版较系统地把牛顿运动定律等经典力学知识介绍到中国。李善兰又和伟烈亚力合译了侯失勒(今译J. 赫歇耳)的《谈天》一书,第一次把万有引力定律及天体力学知识介绍到中国。不久,李善兰又译了奈端的《数理格致》(即牛顿的《自然哲学的数学原理》)一书的前3卷,后因故中断。1867年,他在南京出版《则古昔斋算学》,汇集了二十多年来在数学、天文学和弹道学等方面的著作,计有《方圆阐幽》《弧矢启秘》《对数探源》《垛积比类》《四元解》《麟德术解》《椭圆正术解》《椭圆新术》《椭圆拾遗》《火器真诀》《对数尖锥变法释》《级数回求》和《天算或问》等13种24卷,共约15万字。他的数学著作,除《则古昔斋算学》外,尚有《考数根法》《粟布演草》《测圆海镜解》《九容图表》,而未刊行者,有《造整数勾股级数法》《开方古义》《群经算学考》《代数难题解》等。

第二章　农业科技典籍翻译概述

◎本章学习目标◎

1. 掌握农业科技典籍译介史
2. 掌握农业科技典籍的语言特点
3. 了解农业科技典籍的翻译方法

农业科技典籍作为科技典籍的一部分，指古代的农业典籍。古代的农业目录主要收录于《中国农业古籍目录》。本书共分为正副两篇，正篇主要介绍我国现存的农业古籍目录，副篇主要介绍中国和世界各地收藏的农书目录。“农业古籍，主要指在传统农业阶段产生的著作，时间从先秦起，直至近代的20世纪前期，农书的内容未受西方现代实验科学影响，基本上讲的仍是传统农业的著作”[①]（张芳，王思明 等，2003）。广义的农业科技典籍主要是指与农业相关的有影响力的典籍，狭义的农业典籍除了耳熟能详的具体的典籍——中国五大农书（《氾胜之书》《齐民要术》《陈旉农书》《王祯农书》《农政全书》），也包括描述了与农业相关内容的典籍，如《诗经》《周礼》《离骚》等。

根据农史学家石声汉[②]的定义，中国农书主要是讲述和总结中国传统农

① 张芳，王思明，等.中国农业古籍目录[M].北京：国家图书馆出版社，2003.

② 石声汉.中国古代农书评介[M].北京：农业出版社，1980.

业技术与生产经验,兼及农业经营管理和农本思想的著作。农业典籍作为科技典籍的重要组成部分,在学术研究中,因着眼点之不同,又被人们称为"农家书""农事书"和"农学书"。农史研究早期,人们多称为"农家书",中期称为"农事书",晚期称为"农学书",称谓的演变也反映出人们对农业典籍认识的逐渐深化。[①]本书所讨论的"农业典籍",指包含农业知识的古代典籍。

第一节　农业科技典籍的译介史

中国农业典籍以种类多、内容丰富深刻而闻名于世。王毓瑚在《中国农学书录》一书中将其按体裁及内容范围分为9大类,分别为综合性农书、月令体农书、通书型农书、天时及耕作专著、各种专谱、桑蚕专书、兽医专书、野菜专著、治蝗书;按内容的地域性划分为全国性农书和地方性农书;按农书的内容属性分为农业通论、农业气象、土壤耕作、农田水利、农具、虫害防治、农作物(粮食、经济作物)、蚕桑等;按撰修者则分为私修农书和官修农书。

农业,古代一般称为农桑,即农耕和植桑养蚕。中国是传统农业大国,重农或农本思想是自古以来的国策与传统。《汉书·景帝纪》云:"其令郡国务劝农桑,益种树,可得衣食物。"《明史·太祖纪一》曰:"遣儒士告谕父老,劝农桑。"正因为中华民族几千年来都很重视农业,重视农业科技,自然产生了大量源于农业生产又反过来指导农业生产的典籍。

中国古代早期的文献中就已经记载了农业科技方面的内容,如甲骨文、金文等古文献。战国时期的诸子百家著作几乎都包含了农业技术方面的讨

① 闫畅,王银泉.中国农业典籍英译研究:现状、问题与对策(2009-2018)[J].燕山大学学报(哲学社会科学版),2019,20(3):49-58.

论,如《吕氏春秋》中的《上农》四篇专门讨论农业。百家思想中还出现了专门强调农业以满足衣食需求的“农家”观点。中国古代早期的文献中就已经记载了农业科技方面的内容,如甲骨文、金文等古文献。汉代出现了《氾胜之书》和《四民月令》,其中《氾胜之书》被公认为我国最早的农书。贾思勰的《齐民要术》是我国现存最早、最完整的农书。陆羽的《茶经》则是我国和世界上最早的茶叶专著。宋代编纂了《授时要录》《大农孝经》等官修农书,民间则出现了现存最早反映江南农业生产的地方性农书《陈旉农书》,以及专门研究蚕桑的农书《蚕经》,还有描绘农业生产过程的《耕织图》。在元朝,司农司编撰了我国现存最早的官修农书《农桑辑要》,出版了第一部兼论南北水旱田生产技术的《王祯农书》。清代的《授时通考》则综合了历代农书的精华。根据任继愈编著的《中国科学技术典籍通汇》(2015年出版),从先秦时期至1840年,中国古代具有代表性的科技典籍共有541部,其中包括43部农业科技典籍。[①](王惠琼,孔令翠,2022)

如植物典籍类《救荒本草》很早就流传到国外,在日本先后刊刻。《救荒本草》在日本德川时代(1603—1867年)曾备受重视,还有多种手抄本问世,当时有关的研究文献达15种。这本书曾由英国药学家伊博恩(Bernard Emms Reed,1887—1949)译成英文。伊博恩在英译本前言中指出,毕施奈德于1851年就已开始研究这本书,并对其中176种植物定了学名。而伊博恩本人除对植物定出学名外,还做了成分分析测定。通过比较,指出《救荒本草》的原版木刻图比《本草纲目》的高明。美国植物学家李德(Reed)在他著的《植物学简史》中也赞颂《救荒本草》配图的精确,并说它超过了当时的欧洲。美国科学史家认为《救荒本草》可能是中世纪最卓越的本草书籍。主要的农书译介如下:

① 王惠琼,孔令翠.20世纪前海外中国农业科技典籍译介研究[J].外国语文,2022,38(3):108-115.

《齐民要术》

《齐民要术》早在唐代就已传入日本。日本宽平年间(889—897年)由藤原佐世编撰的《日本国见在书目》中就已涉及《齐民要术》。后经约19世纪法国来华耶稣会士传到欧洲,英国学者达尔文(1809—1882年)在其名著《物种起源》和《植物和动物在家养下的变异》曾提及参阅了"一部中国古代百科全书",并援引有关事例作为进化论的佐证,有说该书正是《齐民要术》。日本现存最早的刻本是北宋天圣年间(1023—1031年)皇家藏书处的崇文院本,被日本当作"国宝"珍藏在京都博物馆中。名古屋市蓬左文库收藏之金泽文库本是现存最早的抄本。

1864年,来华法国汉学家兼农业专家西蒙(Eugene Simon)对《中国纪要》第11卷中《中国的绵羊》一文重新加注,刊于巴黎《风土适应能力学会会报》。西蒙在论及中国农业的论文中也介绍了《齐民要术》并倍加称颂。严格来说,来华传教士和汉学家对《齐民要术》的传播并非真正意义上的译介,而是以该书为纲或援引书中内容撰写的报道或文章。

1958年,石声汉教授历时三年多校注的《齐民要术今释》出版。该书甫一刊出便蜚声中外,得到英国著名中国科学技术史专家李约瑟博士(Joseph Terence Montgomery Needham)等人的关注。是年,石声汉自译的英文版本 A Preliminary Survey of the Book: *CHI MIN YAO SHU*(《齐民要术概论》)问世。译本概要性介绍了《齐民要术》,但仅翻译了书中部分内容,这部农学史上的巨著至今仍无英文全译本面世。①

1945年以来,经国内外学者的译介,《齐民要术》在国际上得到进一步传播。在日本和欧美一些国家,对《齐民要术》的研究也很流行,并称为"贾学"②③。

---

① 王翠.论新时代中国农学典籍的翻译与传播[J].南京工程学院学报(社会科学版),2019,19(4):18-23.

② 郭超,夏于全.传世名著百部[M].北京:蓝天出版社,1998.

③ 张保国.中国农学典籍《齐民要术》译介研究述评[J].乐山师范学院学报,2020,35(3):46-51.

《茶经》

《茶经》的首个英译本是由 William Harrison Ukers（1873—1945）翻译，题为 *All about Tea*，于是 1935 年在美国由 The Tea and Coffee Trade Journal Company 出版,并于 2007 年再版。不过,这是一个节译本。1974 年,英国出版了 Francis Ross Carpenter 翻译的全译本 *The Classic of Tea*，并于 1995 年重印,收录于《大百科全书》,且配有精美图片,文字通俗易通。由国内学者姜欣、姜怡翻译的 *The Classic of Tea* 被列入汉英对照《大中华文库》,于 2009 年由湖南人民出版社出版。这是第一部由中国人自己翻译的《茶经》英语全译本。①

《茶经》于南宋时期传入日本。12 世纪中期,僧人荣西两次来华,将《茶经》手抄本带回日本。江户时代日本开始了对《茶经》的翻刻。1774 年,大典禅师对《茶经》加以训点,并用片假名混杂汉字详注,撰写了《茶经详说》。

《茶经》流传至欧洲的时间相对较晚,但传入后也被陆续译为英、德、法、意等多种西语文字。意大利是较早研究中国茶的欧洲国家。1559 年,意大利著名作家詹巴迪斯塔·拉摩晓出版的《茶之摘记》《中国茶摘记》《旅行札记》等著作都有对《茶经》的记载。当代威尼斯学者马克·塞雷萨（Marco Ceresa）于 1991 年出版了《茶经》的意大利语译本,这个译本是目前西方最全的《茶经》译本,其第一版面市后很快即告售罄,可见欧洲社会对《茶经》的热情之高。②

《农政全书》

《农政全书》在 17 世纪中叶开始流传到日本。日本江户时代著名学者中村惕斋在 1666 年发表的《训蒙图汇》所列参考文献中提及《农政全书》。

---

① 王宏,刘性峰.国内近十年《茶经》英译研究(2008—2017)[J].外文研究,2018,6(2):64-69,108.

② 袁梦瑶,董晓波.《茶经》译介推动中国茶文化走向世界[N].中国社会科学报,2019-03-15.

历史上的日本经常发生饥荒，江户时代总共有154次饥荒，其中21次范围甚广且严重（吉武成美，1982）。包含有救荒内容的《农政全书》传入日本后，本草学大师松冈玄达等立刻从《农政全书》中析出《救荒本草》及《野菜谱》并加注后在京都刊行。一批日本农学家据此总结了一些可食用植物的利用方法，为赈济饥荒发挥了较好的作用。

《农政全书》传入朝鲜后，思想家朴趾源在1783年写成的《热河日记》的《车制》中不但提到了《农政全书》，还提到了《天工开物》和《耕织图》等，在所著《课农抄》中也多次引用。日本和朝鲜还有多位学者译介了《农政全书》，本书将在下文讨论《天工开物》时一并讨论。1870年，意大利人安德烈奥将《农政全书·荒政》篇中的"玄扈先生除蝗疏"部分内容翻译为意大利文，题目为《论蝗虫〈农政全书〉论述提要》。

"《农政全书》最迟在18世纪传到欧洲。在19世纪，该书收藏于法、英、德、俄、荷等欧洲各国及美国的大图书馆里，是欧洲人最先注意的中国科技著作之一。1735年，在巴黎用法文出版了一部四卷本的《中华帝国全志》，其中卷二转载了《农政全书》卷31—39——《蚕桑》篇的法文摘译，这是此书译成欧洲语之端始。《中华帝国全志》是巴黎耶稣会士杜赫德（Jean-Baptiste Du Halde，1674—1743）根据在华的27名教士的稿件编辑而成的，是有关中国的百科全书式的名著。《农政全书》的法文摘译者是殷弘绪。殷弘绪字继宗，原文名弗朗索瓦萨维尔·丹特拉格尔（Francois-Xavier D'Entrecolles，1674—1741）。"①

《中国科学技术史》首次全面英译介绍了《农政全书》。"《中国科学技术史》是英国著名中国科学技术史研究专家李约瑟博士耗时50余年，组织世界各地该领域专家撰写的一部中国科学和技术历史的鸿篇巨制。该书在丰富的史料和大量的实地调查基础上，深入系统地介绍了中国古代科学和技

① 潘吉星.徐光启著《农政全书》在国外的传播[J].情报学刊，1984(3)：94-96.

术取得的辉煌成就,为西方人公正了解和认识中国古代科学和技术发挥了重要作用,堪称中西文化交流史上的一座丰碑。《中国科学技术史》共7卷。第1卷为《导论》;第2卷为《科学思想史》;第3卷为《数学与天文学、地球科学》;第4卷为《物理学和物理技术》;第5卷为《化学与化工技术》;第6卷为《生物与生物技术》;第7卷为《社会背景》。《中国科学技术史》第6卷《生物与生物技术》分为11册,其中第2册为《农业》,1984年出版,撰稿人是英国著名中国科技史研究专家白馥兰教授(Francesca Bray)。该册第64页至70页首次用英语比较全面地介绍了《农政全书》。"[①]

《天工开物》

"《天工开物》于17世纪首先传入日本。日本本草学家、农学家都高度重视,佐藤信渊还依据天工开物思想提出了富国济民的'开物之学'[②](潘吉星,2013)。1952年,京都大学人文科学研究所首次将《天工开物》全译成日语,该译本为《天工开物》第一个外文全译本。1969年,薮内清译出了第二个日文全译本。18世纪末,《天工开物》经日本传到朝鲜。朝鲜李朝后期一些实学派学者也很重视。1783年,朴趾源在其游记《热河日记》里第一次向朝鲜推介和传播了宋应星和他的《天工开物》。徐有榘著《林园经济十六志》与李圭景著《五洲书种博物考辨》和《五洲衍文长笺散稿》等都多次引用《天工开物》。到了1997年,韩国汉城外国语大学崔炷翻译后由传统文化社出版"。[③] "《天工开物》在欧美国家中最早传到法国。1830年,法兰西学院汉学家儒莲(Stanislas Julien,1793—1873)率先将《丹青》中论银朱部分译成法文。此后的约10年时间,《天工开物》被转译为多种语言。19世纪上半叶,《天工开物》引起了法兰西学院儒莲的注意。1830年,他首次把《天工开物·

① 李海军.18世纪以来《农政全书》在英语世界译介与传播简论[J].燕山大学学报(哲学社会科学版),2017,18(6):33-37,43.

② 潘吉星.《天工开物》在国外的传播和影响[N].北京日报,2013-01-29.

③ 孔令翠,刘芹利.中国农学典籍译介梳理与简析[J].当代外语研究,2019(4):106-114.

丹青》部分译成法文。1833年，又在《化学年鉴》上发表《五金》卷中关于制铜的译文。1837年，又把《蚕桑篇》摘译成法文。19世纪《天工开物》第一卷被译成法文后，即被转译成其他欧洲语。现在已被全译成日文和英文，部分被译成德、俄、意大利文。该书甚至被一些国家称为'技术百科全书'。1966年，伦敦和宾夕法尼亚的大学出版社翻译出版了该书。1969年，日本东京平凡社又据1952年版本重版本书"。[①]"到了20世纪的1964年，德国洪堡大学研究生蒂洛把《天工开物》与农业有关的《乃粒》《乃服》《彰施》及《粹精》译成德文并加了注释还作专题研究，取名为《宋应星著前四章》"。[②]

目前，《天工开物》有三个英译本。1966年，《天工开物》第一个英文全译本由美国宾夕法尼亚大学华裔学者孙守全、任以都夫妇翻译，书名为 *Tien kung K'ai-wu : Chinese Technology in the Seventeenth Century*。1980年，李熙谋和李乔苹等主译的《天工开物》英文全译本在中国台湾出版，书名为 *Exploitation of the work of nature-Chinese Agriculture and Technology in the* XⅦ *Century by Sung Yingsing*。2011年，广东教育出版社出版了《大中华文库》《天工开物》汉英对照本，由王义静、王海燕和刘迎春英译，书名为《天工开物》(*Tian Gong Kai Wu*)。

第二节　农业科技典籍的文本特点

一、农业典籍中的哲学智慧

农业典籍不仅对于了解中国传统农业文化有重要的意义，同时也为翻译研究和跨文化交流提供了新的视角和范例，可帮助读者更好地认识和保

① 费振玠，曹洸．从《天工开物》外译情况谈科技翻译[J]．上海科技翻译，1988(2)：41-43.

② 袁梦瑶，董晓波．《茶经》译介推动中国茶文化走向世界[N]．中国社会科学报，2019-03-15.

护传统文化,促进中国传统文化的传承与发展。对于研究农业典籍文化对外传播与更深入地了解中国农业文化的丰富内涵,感受传统文化的魅力与智慧,具有指导作用,为农业典籍中的哲学智慧的对外传播提供了参考和指导。

中国农业典籍是以农业技术传播文化。农业文化蕴含了中国上下五千年的历史文明与传承。中国农业典籍中可见中国文化的哲学智慧,古代人民的农业智慧在农业典籍中有根可寻,如我国油脂文化可见于农业典籍的记载。从《考工记》《氾胜之书》《梦溪笔谈》《陈旉农书》《农桑辑要》《农政全书》《王祯全书》和《天工开物》等农业典籍考究了油脂文化的史料来源。农业典籍蕴涵农业文化哲学智慧,如生态智慧、人与自然和谐共生的智慧、传承经验的智慧、"天人合一"的智慧等。《天工开物》是我国古代较完整、较全面的制油典籍,反映了三百多年前劳动人民的制油技术水平。《齐民要术》的作者是东汉时期的贾思勰。书中强调,"山、泽有异宜。山田种强苗,以避风霜;泽田种弱苗,以求华实也。顺天时,量地利,则用力少而成功多。任情返道,劳而无获"[①](贾思勰,2015)。要"顺天时,从物宜",要遵循自然规律去耕作,这就很好地体现出了天人合一的思想。《齐民要术》被誉为"中国古代农业百科全书",包括了油料作物的选种、播种、施肥、管理等方面的知识,以及油料的采收、储存和加工等技术。农业典籍注重实践经验的总结和传承正是中国哲学中实践经验性的重要体现。通过记录和传授农业技术、经验和管理知识,保证了农业技术的传承和发展。中国传统哲学中的"仁爱"和"和谐"观念,即强调社会和谐、人际关系的和睦在典籍的农业生产生活记载中可见。

《农政全书》卷二十三记载:"[油榨]取油具也。用坚大四木,各围可五尺。长可丈馀,叠作卧枋于地。其上作槽,其下用厚板嵌作底槃,槃上圆凿

① 王惠琼,孔令翠.20世纪前海外中国农业科技典籍译介研究[J].外国语文,2022,38(3):108-115.

小沟，下通槽口，以备注油于器。凡欲造油，先用大镬炒芝麻，既熟即用碓舂，或辗碾令烂上甑蒸过，理草为衣贮之，圈内累积在槽。横用枋桯相栘，复竖插长楔，高处举碓或推击榨之极。”这句话描述了榨油工具的制作材料和结构，使用坚固的木材制作工具，可以增加工具的耐用性并延长使用寿命，减少对自然资源的浪费。同时提到了使用传统的炒、碾、蒸等方式来加工芝麻，强调了传统的农业生产方式和对原材料的合理利用。“复竖插长楔，高处举碓或推击榨之极”描述了使用竖插长楔固定碓，以便在高处举碓或推击来磨碎芝麻。这种方式可以减少人力劳动，提高效率，是一种生态农业中的既节约资源又提高效率的方法。

由此可见，农业典籍中蕴藏的生态农业哲学智慧意义重大，彰显了我国传统农业文化的博大精深。这些智慧不仅是文化强国建设的基石，也是其根本所在。挖掘农业典籍中的哲学文化，以有效有序的方式传播中国农业典籍文化，进一步构建中国农业传统文化的经典外宣形象，同时也加强了文化强国的外宣建设。这一途径将有助于提升中国农业文化的国际影响力，推动传统文化的传承与创新，为我们的强国文化注入新的思想和魅力。

农业典籍为农业生态学提供了丰富的思想资源。典籍中的许多原则和实践至今仍然适用，对我们建立和谐的农业生态文化系统提供了有力的理论支持。文化强国背景下如何传播中国生态农业文化，讲好中国故事，农业典籍对中国油脂文化的传播起着重要的作用，通过对中国油脂文化的深入研究和传播路径的探讨，有助于我们更好地传承和发展农业典籍中的文化哲学智慧。农业文化哲学对外传播路径主要有以下几个方面：

第一，加强农业典籍的翻译出版与翻译研究。农业典籍蕴含中国传统农业的精髓，是古代劳动人民智慧的结晶，是具有中国哲学智慧的农业文化，历史悠久，彰显了劳动人民的“天人合一、顺应自然”的农业文化思想特色。目前,关于农业典籍外译研究较少,未形成系统性典籍翻译模式,亟待学者开拓新的研究思路,为我国农业典籍的高质量翻译提供参考。

第二,传播“和谐”生态农业文化,促进我国农业文化国际化,有利于国际有机农业发展,建立农业典籍的对外传播平台,通过多维角度宣传可促进世界了解中国传统生态农业文化,增进不同国家和民族之间的友谊和合作。我国的农业文化是历经数千年积淀形成的,其中包含了丰富多彩的哲学思想和道德观念。农业典籍所包含的农业文化融入自然,弘扬和谐与包容,是具有深厚人文情怀和社会价值的“和谐”生态农业文化。

第三,加强传统“顺应”“农本”与“耕本”的生态农业文化宣传,将具有中国哲学精髓的农业文化传播给世界。这不仅对于推广我国环保、生态、可持续发展的农业模式有着重要意义,而且有助于农业生态文化在西方世界的宣传,促进文化强国的建设,保护和传承我国优秀的传统文化,推动文化创新和创造,培育具有国际影响力的文化产业,提高国家的文化软实力和国际形象。通过文化强国建设,提升我国的文化自信,凝聚民族精神,增强国家凝聚力,推动全球农业生态文化一体化发展。

第四,建立全球电子化农业博物馆,实现跨文化交流,有利于全球农业生态文化共同体建设。建立开放式全球平台,以中国哲学智慧故事为主线,展示古代农业文明与文化。通过文字、图片、音频、视频等多种形式,讲述故事,介绍农业技术、农民生活、农业智慧等,向全球观众展示中国的传统生态农业文化和油脂文化,让世界更好地了解中国农业文化的深厚哲学底蕴。

第五,建立产教融一体化机制,开创特色农业典籍翻译教育,培养农业典籍翻译人才,加强关于农业典籍翻译的教材编写,融入中国农业哲学的基本概念、原理和实践经验,将中国农业哲学与翻译教育或对外传播教育联系起来,有利于农业典籍哲学思想的对外传播。

总之，在传播中国农业生态文化哲学智慧时，需要克服文化差异等挑战。中国农业典籍的哲学思想是基于中国传统文化的，与其他国家和地区的文化存在一定的差异，在翻译中可通过加注农业哲学文化知识点，帮助读者理解典籍中的专业术语和文化背景。

二、农业科技典籍文本的语言特点

首先，这些文本中使用了大量专业术语，涵盖了农业生产的各个领域，如作物栽培、畜牧养殖、土壤肥力等。这些术语丰富而精确，有助于准确传达知识和技术。其次，农业科技典籍文本通常采用简练的语言表达，突出重点，避免冗长的描述。这种简练的语言风格使读者能够迅速理解和掌握知识，提高学习效率。此外，农业科技典籍文本具有较强的系统性。它们按照一定的体系结构组织知识，将相关内容有机地连接在一起，形成完整的知识体系。这种系统性有助于读者系统性地学习和应用知识。农业科技典籍文本在语言表达上通常较为正式和严谨。它们避免夸张和夸大，注重客观性和准确性，确保内容的可信度和权威性。最后，农业科技典籍文本注重实用性。它们围绕农业生产实践展开，提供实用的技术方法和经验，帮助读者解决实际问题，具有一定的实用性和指导性。

（一）词汇方面

农业典籍涵盖了广泛的农业知识，包括耕作、畜牧、园艺、林业等各个方面，为农业实践提供全面的指导和信息。这些典籍记录了丰富的农业实践经验和技术，旨在帮助农民解决实际问题，提高农业生产效率。这些典籍承载着丰富的农业传统和文化，反映了古代农业技术的发展历程，是农业文化的珍贵遗产。尽管古代的农业典籍可能缺乏现代科学技术的支持，但它们在当时代表了最前沿的农业知识和理论，为后世农业科技的发展奠定了基础。这些典籍对后世农业文献和农业实践产生了深远的影响,成为后来农业科技发展的重要参考和基础。通过研究和理解农业科技典籍的特点,我们可以更好地了解古代农业技术和文化,同时也可以借鉴其中的经验和智慧,促进当代农业科技的发展和进步。

例如,《茶经》语言具有浓郁的色彩美。作者除了直接采用准确的颜色词来描述茶色,还采用了一系列暗含色彩意义的词语,将茶的各种色泽通过隐喻的语言形式表现出来。《茶经·五之煮》中有一段关于茶汤精华"沫饽"的描写:"华之薄者曰沫,厚者曰饽,细轻者曰花,如枣花漂漂然于环池之上,又如回潭曲渚青萍之始生;又如晴天爽朗有浮云鳞然。其沫者,若绿钱浮于水渭,又如菊英堕于鐏俎之中。饽者,以滓煮之。及沸,则重华累沫,皤皤然若积雪耳。"

《茶经》通过对具体物象的描写将茶叶品鉴行为上升到更高的审美境界。恰当生动的具象引领读者进入想象空间,形成无限美的意象。《茶经·五之煮》:"其沸,如鱼目,微有声,为一沸;缘边如涌泉连珠,为二沸;腾波鼓浪,为三沸。"作者用"鱼目""涌泉""腾波鼓浪"三个具体的物象描绘了煮茶时沸腾的不同状态:开始沸腾时是如同鱼眼般小小的气泡,开合之间发出微微的声响;接着煮,锅边冒出像涌泉一样的水泡,有咕噜之声;再煮,沸腾的水像翻腾的波浪,声音又大又急。文本虽然不是写声音,但形象生动的具体物象描写仿佛使人感受到煮茶之时水沸腾的样子,声音由微弱变得强烈,煮茶状态呈现出由静到动的美感。《茶经》语言的形式美体现在两方面:一是用词简洁干净,言简意赅;二是大量运用对比、比喻、排比等修辞方式。这些修辞方式的使用不仅将茶叶品评的标准、技艺生动地描述出来,更使文章具有一种行文结构上整齐划一的美感,呈现出内在和谐的行文气韵。①

农业词汇具有独特的语言特点。首先,这些词汇通常是专业术语,用于

① 王秀丽.《茶经》语言的审美特质[J].农业考古,2018(2):170-172.

描述农业实践、技术和科学知识，具有一定的专业性和技术性。其次，许多农业词汇源自古代文献和传统农业实践，反映了古代农业文化和技术的特点，可能存在一些古老的用词和表达方式。此外，农业词汇具有实用性，能够准确描述不同的农业工具、作物、动物等，反映了农业领域的实践性和操作性。随着农业科技的发展，现代农业词汇越来越强调科学性和准确性，体现了农业科技领域的专业化和现代化。另外，农业词汇也具有地域性和多样性，反映了不同地区的农业环境和文化背景。最后，一些农业词汇可能源自民间传统和口头传承，体现了农民对土地和农业生产的理解和认识。通过了解这些语言特点，我们可以更好地理解农业领域的专业术语和文化内涵，促进农业领域的交流和发展。

如学者韩忠治指出"《农政全书》的语言以实用为目的，力避怪僻和浮华，加之作者总结、推广农业技术和农业生产经验的目的所决定，全书的语言必定是通俗易懂、接近当时口语的；而在'农政'和征引前代农学著述部分，则主要是向统治者、士大夫提出农业生产的政治措施，风格相对典雅。这种雅俗兼具的语言风格，既不同于'台阁体'的萎靡，又不同于'前后七子'的复古，是明代真实语言面貌的记录。内容方面，该书无愧于'全书'的称谓，60卷的宏大规模，内容涉及历代农业政策和农业经济，尤其是谷物、蔬菜、竹木的种植和管理，植物病虫害防治，畜牧、兽医、养鱼、水利、荒政等方面，所用术语或来自民间，或由古语演化，全面反映了古白话俗语词土生土长的'大农业'用语的面貌和轮廓。作者及著作年代确凿，加之成书距今不过三百多年，完全保持了原书面貌。和其他农书语言一样，《农政全书》的语言也具有专业性、口语性的特点，同时也具有新颖性、方言杂糅的特点"①。

① 韩忠治.《农政全书》词汇研究[D].河北师范大学，2015.

(二)句法方面

农业典籍因其科学性和实用性,句式多样,既简洁明了,又常用复杂句式。书中大量使用古汉语结构,并列句式以整合相关信息,阐述农业理念和政策时融入比喻和象征手法。为强调关键农业原则或技术,常采用重复句式加深印象。在讨论农业问题及解决方案时,多用条件句和假设句展示应对策略。作为指导性著作,农业典籍在提出建议时常用命令或劝告语气,引导实际操作。

例如,《农政全书》在介绍木棉种植方法时,原文直接以"孟祺《农桑辑要》曰:择两和不下湿肥地,于正月地气透时,深耕三遍,耙盖调熟,然后作成畦畛。……"为原文的开头,体现了简洁明了、指导性强等特点。

《汉礼仪志》:"皇后祀先蚕,礼以中牢。魏黄初中,置坛于北郊,依周典也。晋置先蚕坛,高一丈,方二丈。四出陛,陛广五尺。皇后至西郊,亲祭、躬桑。北齐先蚕坛,高五尺,方二丈,四陛,陛各五尺。外兆四十步。面开一门。皇后升坛,祭毕而桑。后周,皇后至先蚕坛,亲飨。隋制:宫北三里,坛高四尺。皇后以太牢制币制祭。唐置:坛在长安宫北苑中,高四尺,周围三十步。皇后并有事于先蚕。其仪,备开元礼。宋用北齐之制,筑坛如中祠礼。"

《通礼义纂》:"后亲享先蚕,贵妃亚献,昭仪终献。夫蚕祭有坛,稽之历代,虽仪制少异,然皆递相沿袭,饩羊不绝。"描述了历代皇后亲自参与蚕祭的仪式,体现了农业(特别是蚕丝业)在国家经济和文化中的核心地位。蚕祭的仪式不仅是对蚕神的敬仰,也是对农业活动重要性的肯定,反映了农业作为国家基础的价值观。通过皇后亲自参与蚕祭,强调了农业的重要性,从而体现了农业为立国之本的思想。

农业典籍以其实用性和指导性为核心特点,其句式结构旨在明确传达农业生产技术的方法与步骤。通过精确的专业术语、复杂的句式结构、浅显易懂的语言，确保了信息的准确性和操作性。此外，农业典籍中还融入了历史典故和哲学思考，不仅帮助读者深入理解农业知识，而且强化了农业作为国家根本的理念，从而推动农业生产的持续发展。通过这种句式特点，农业典籍有效地将理论与实践相结合，为读者提供了既具体又系统的农业指导。

三、农业科技典籍文本的结构特点

农业科技典籍文本的结构特点主要体现在其分章节结构、渐进式展开、实例分析、图表配合和总结回顾等方面。首先，农业科技典籍通常按照不同的主题或内容进行分章节编排，使读者能够系统地学习和掌握相关知识。其次，典籍文本常采用渐进式的方式展开内容，从基础知识开始逐步深入，帮助读者循序渐进地理解和掌握农业科技知识。此外，为了帮助读者更好地理解和应用知识，典籍文本通常会通过实例分析或案例说明的方式，具体展示技术应用过程和效果。同时，图表配合也是常见的手段，以图文结合的方式展示数据和技术要点，增强读者对内容的理解和记忆。总体而言，农业科技典籍文本的结构特点旨在帮助读者系统地学习和掌握农业科技知识，提高农业生产水平。

"农业英语的文体特征鲜明，常用无人称词汇和专业术语。其中，专业术语指局限于表达本学科概念的准确、狭义的词汇，这些词汇在日常文本中很少出现，因而会给读者带来很大的理解困难，如扦插(cuttage)、亲本(parent)、根冠(cap)、粟/黍(sorghum)等。另外，中国传统农业文化孕育了一批表达独特和蕴含特殊含义的专有词汇，这些词汇很难在其他语言里找到完全对应的表达，如'负阴抱阳''顺耕逆耕''刀耕火种''铁犁牛耕'以及一些农谚、口诀等，在翻译过程中很容易出现词汇空缺，常常需要译者在充分理解原文的基础上，综合使用多种方法来进行翻译，这大大增加了农业典籍英译的难度。此外，农业英语词汇通常具有一词多义的特点。例如，当提到施肥技术时,人们最先想到的英文对应单词可能是‘fertilizer’,其实,中国农业生产中的汉字‘肥’在用英语表达时有很多种译法:base manure(底肥)、top-dressing(追肥),straw manure(草肥)、ash fertilizer(灰肥)等。一个‘肥’字,在英语中多达数十种表达形式,正所谓异词同义,令人眩目。”①

对于《黄帝内经》的翻译,“语言国情学认为,凡是含有国情的概念均应音译,以利于保持其内涵。中医基本理论中的核心概念均含有国情,如阴阳、五行、脏腑、精、气、神等。这些概念在英语语言中基本上没有完全对应的说法,翻译时无论直译还是意译都无法完全表达清楚原文的内涵。如‘精’现在一般译作essence,‘神’一般译作spirit或mind,这种译法其实只表达了中文概念的部分内涵。对于诸如此类的概念,我们在翻译时均予以音译,另以括号形式将现行译法作为一种文内注解并附于有关音译概念之后,以帮助读者理解。如‘精’译作Jing (essence),‘神’译作Shen (spirit),‘五行’译作Wuxing (five elements)。译文自始至终采用这种音译加文内注解的译法,表面上看似重复累赘,实则不断向读者传递来自远古的原本信息,使读者明白括号中的注解只是一种辅助解读手段或该概念的表面之意,而非其实际含义。”②

农业典籍的文本特点和语言特点在具体例子中得到了充分体现。例如,《农政全书》中记载的《四时类要》中提到种植桑的方法:“种桑,土不得厚,厚即不生。待高一尺,又上粪土一遍。③”这句话简洁明了地表达了种植桑树时土壤的处理原则,即土壤不宜过厚,需要等待桑树长到一定高度后再施粪土,使读者能够清晰地理解操作步骤。此处的语言特点包括古文表达

---

① 闫畅,王银泉.中国农业典籍英译研究:现状、问题与对策(2009-2018)[J].燕山大学学报(哲学社会科学版),2019,20(3):49-58.

② Yellow Emperor's Canon of Medicine: Plain Conversation. Vol. Ⅰ, Ⅱ, Ⅲ. 黄帝内经·素问(全三册)[M]. 李照国,刘希茹,译.西安:世界图书出版公司,2005.

③ 徐光启.农政全书[M].上海:上海古籍出版社,2010.

和术语精准，体现了农业典籍的语言风格。

学者潘吉星在《天工开物》的前言描述了其特点："第一，《天工开物》不是历代文献的堆积，而是据生产现场的实地调查著述而成。宋应星通过实地调查，详细记述了生产领域的技术过程、操作要点、原料及产品、生产工具，以及先进的科技成果。除了文字表述，还用插图将生产情景再现出来。第二，《天工开物》是在一种先进而又有特色的技术哲学思想的指导下写成的，我们将其概括为'天工开物思想'。这种思想强调人与天相协调、人工（人力）与天工（自然力）相配合，通过技术从自然界中开发出有用之物。第三，《天工开物》对原料与能源的消耗、成品产率、设备构造及各部件尺寸等，都尽可能给以定量的描述，且绘出工艺操作图，在某种程度上好像是近代科学家对传统技术写出的调查报告。西方近代科学以其数学化而与中世纪诀别，《天工开物》在这方面走得相当远。生产过程中涉及的长宽高、重量、容积比率、时间等技术指标都作了描述，其中长度精密到分寸、重量精密到钱这样的数量级。书中的大量设备图有立体感，各部件长短协调，有如工程画；画面上人物操作逼真、表情自然，联起来好像中国古代技术史的长卷画面。"①

另外，《务本新书》中描述种植椹的方法也是一个很好的例子："四月种椹，东西掘畦，熟粪和土耧平，下水。水宜湿透，然后布子。②"这段文字详细描述了种植椹的时间、方法和注意事项，包括掘畦、施肥、浇水等步骤，使读者能够清晰地了解种植椹的操作流程。这里的语言特点包括细致描述和精准术语，有助于读者准确掌握种植椹的技术要点。

① Tian Gong Kai Wu 天工开物[M]. 王义静，王海燕，刘迎春，等译. 广州：广东教育出版社，2011.

② 贾思勰. 齐民要术今释[M]. 石声汉，校. 北京：中华书局，2009.

第三节 农业科技典籍的翻译过程

方梦之在《译学辞典》中提到“翻译过程(process of translation)指翻译活动所经过的程序,一般认为包括三个阶段:理解原文,用目的语表达、校验修改译文。其中,理解是表达的基础或前提,表达是理解的结果。奈达提出的过程是:1.分析——从语法和语义两方面对原文的信息进行分析;2.传译——把经过分析的信息在脑子里从原语转换成译语;3.重组——把传递过来的信息重组成符合要求的译语;4.检验——对比原文意义与译语意义是否对等。总的来说,翻译过程主要是理解与表达的过程,即认识与实践、分析与综合的过程,它们既是同一过程彼此衔接,又是互有交叉、互臻完善的两个阶段,是同一个问题不可分割的两个侧面。”①

农业科技典籍的核心翻译过程主要包括两个方面:理解阶段和表达阶段,总的来说,一般的翻译都包括理解跟表达两个部分。“理解(comprehension)指弄懂原语的种种含义,是译者认识事物之间联系的本质与规律的一种思维活动。理解分为直接理解和间接理解。直接理解是马上实现的,是过去已经理解的事物的重现。间接理解是逐步实现的,需要从不理解到理解的过渡过程,通常是在感知、表象与再造想象的基础上借助于思维过程实现。从语言是客观世界的能指这一角度来说,任何语句序列都是对客观世界的描述,它可能是具体的,也可能是抽象的,那么译者必须把语言的所指找出来,搞清楚原作者要表达的语气、文体、语用含义等。由于原文的内容不同,理解的性质也不同。可以区分为:1.对语言的理解;2.对事物意义的理解;3.对事物类属的理解;4.对因果关系的理解;5.对逻辑关系的理解;6.对事物内部

① 方梦之.译学辞典[W].上海:上海外语教育出版社,2004.

构成、组织的理解。阅读中的理解，通常经历以下阶段：1.字面的理解水平；2.解释的水平；3.批判性阅读；4.创造性阅读。因此，不能仅将揭露事物本质与规律视为理解。理解以旧经验、旧知识为基础。经验的丰富性与正确性，已获得的基本知识的数量与质量、思维发展水平等，都影响理解知识的水平。对于翻译来说，理解是关键，只有正确的、深刻的理解，才可能有完善的表达。”①

“奈达(Nida 1969)借用乔姆斯基的转换生成语法理论，提出逆转换模式，将翻译分成三个阶段:分析(analysis)、转换(transfer)和重构(restructuring)；贝尔(Bell 1991)从句法、语义和语用等方面对源语和译语中的信息过程进行研究，他认为翻译分两个阶段进行:源语语篇分解(analysis)和译语综合(synthesis)；斯坦纳(Steiner 2001)把翻译分成四步：信赖(trust)、侵入(aggression)、吸收(incorporation)和补偿(restitution)。”②

理解阶段包括三个要点：1.理解全文，推敲词义。2.读懂语法，理清句子关系。3.全局解读原文。对于农业科技典籍的翻译，不同于一般性翻译，需要译者具备农业科学知识和文学素养，因此翻译时理解是第一步。第一步对于科技词汇的理解尤为重要。判断词义不仅要根据词汇搭配来判断，还要结合上下文来推敲。对于科学性、逻辑性很强的科技英语翻译，即便是在同一个句子里，同一个词也会具有不同词义。例如，本草采用中文原文、中文注释、英文译文、英文注释进行编排。个别中药名经多方考证仍无法确定其拉丁名称，翻译成拉丁文的时候，采取Materia Medica加音译的方法，如“船底苔”译为Materia Medica Chuanditai。本草名称的翻译采取“四保险”的翻译方法，即每个本草名称均按拼音、汉字、英文和拉丁文的方式进行翻译，如“萹竹”译为Bianzhu。本草名称如果是三个字及以下，其音译合并在一

① 方梦之.译学辞典[W].上海：上海外语教育出版社，2004.

② 王小凤，张沉香.科技英语翻译过程的多维思索[J].中国科技翻译，2006(4)：33-36.

起;如果是四个字及以上,根据文义将其音译分开,便于阅读,如“鹅不食草”音译为Ebushi Cao。古籍名称采用音译的方法翻译,括号中附以中文和英文翻译,音译中的每个字独立音译。如《古文尚书》译为Gu Wen Shang Shu(《古文尚书》,The Chinese Ancient Classic)。①

农业科技典籍翻译过程的最后一个阶段是“检验(test)是奈达翻译过程四阶段[另三个为:分析(analysis)、转换(transfer),重组(restructuring)]中的最后阶段。即将译文与原文进行比较,对照,检查译文跟原文在文字上或语法上是否对应,而更重要的是译文意义跟原文意义(包括文体风格)是否对应”②。

第四节 农业科技典籍的翻译策略

“按照文本类型来分,农业典籍属于信息型文本,蕴含着中国传统文化和哲学思想。以归化和异化的选择为例,译者若在翻译过程选择了归化策略,就需要变动原文的体裁、文化意象、语言结构等元素,努力使译文向目的语国家的语言习惯和文化价值观靠拢,从而实现真正意义上的文化交流;而异化策略则要求译文尽可能保留原有的风味,不改变原文语义、结构、文化意象等,这有助于目标读者直接,准确地解读异域文化,从而实现不同文化间的交流与碰撞。而对于一些文化负载词、特色浓郁词汇或文化典故,由于两种语言文化之间的鸿沟较大,单纯地异化将难以达到有效的交流目的,在这种情况下,为了使目标读者更好地理解,译者也会综合使用直译加注释等方式进行处理。”③

“在翻译《黄帝内经》时,我们确定的一个基本原则就是‘译古如古,文不

① 张志聪.本草崇原[M].孙慧,译.苏州:苏州大学出版社,2021.

② 孔令翠,刘芹利.中国农学典籍译介梳理与简析[J].当代外语研究,2019(4):106-114.

③ 王宏,刘性峰.国内近十年《茶经》英译研究(2008-2017)[J].外文研究,2018,6(2):64-69,108.

加饰'。就方法而言，基本概念的翻译以音译为主。释译为辅，篇章的翻译以直译为主、意译为辅。以此法翻译之译文，读起来虽不十分流畅，但却能最大限度地保持原作的写作风格、思维方式和主旨（李照国）。"①

"在文章的翻译上，我们原则上保留原文的结构形式和表达方式，译文中尽量不增加词语。但由于古汉语的表达特别简洁，翻译时若不增加字词，有时很难使一句话结构完整。在这种情况下我们也适当增加一些词语，以便使一句话结构完整，表达流畅。像这样一些为句法结构的需要或为语义表达的需要而增加的词语，译文均置于[]之中。译文中频繁出现[ ]，的确有碍观瞻，但惟有如此方能保持原文的本意，使读者理解何为原经文之语，何为注解之语，从而防止衍文的出现（李照国）。"②

"为了更准确地展现和传递《救荒本草》的基本信息，本书的本草采用中文原文、中文注释、英文译文、英文注释予以编排。本草名称的翻译采取'四保险'的翻译方法，即每个本草名称均按拼音、汉字、英文和拉丁文的方式进行翻译，如'荠菜'译为 Jicai[荠菜，shepherd' spurse ，Capsellae Bursa-Pastoris (L.)Medic.]。本草名称如果是三个字及以下，其音译合并在一起；如果是四个字及以上，根据文义将其音译分开，便于阅读。古籍名称采用音译的方法翻译，括号中附以中文和英文翻译，音译中的每个字独立音译。例如，《图经本草》译为 Tu Jing Ben Cao [《图经本草》，Illustrated Classic of Materia Medica]。"③

"《山海经》的译者安妮·比勒尔将所有的人名、地名、动植物名和神话人物等全部根据其本意翻译成对应的英文，而不是采取音译的办法。比勒尔称：'采用意译法译人名地名等能在文中避免冗长和难懂的音译并立刻使译文充满生气。'该译本的另一特点是将所有动植物名词均对应译为盎格鲁-

①② 李照国译《黄帝内经》前言.

③ 朱橚.《救荒本草》汉英对照[M].范延妮，译.苏州：苏州大学出版社，2019.

撒克逊词汇,而不是充满学究气的拉丁词语。"①

王宏在翻译大中华文库的《山海经》提到:"鉴于列入《大中华文库》的《山海经》英译全译本的读者对象主要是英美国家的普通读者,我们在翻译此书时制定的总原则是,译文要做到'明白、通畅、简洁'。'明白'指所译出的译文要让普通读者看得懂。'通畅'指译文本身不能过度拘泥于原文结构,造成行文梗阻,阅读吃力。'简洁'之所以排在第三,是因为我们要求的译文'简洁'是建立在'明白、通畅'基础上的。如果只是片面追求译文的'简洁',以牺牲译文的'明白、通畅'为代价,就不宜效仿。为方便西方读者理解,我们对所遇到的疑难之处均进行了认真考辨,在语言转换、内容表达等方面尝试了各种翻译技巧,最终形成了我们自己的译本。具体来说,技术性强的条目尽可能使用简洁易懂的语言,采用解释性译法。而对叙事性条目则采用直译法,尽可能再现原文风格。《山海经》原文涉及的人名、地名和人物众多,为了方便读者辨认,我们尽量采取音译法,以求统一。对中药和植物的译名,我们尽量采用的是通用名称(common names),而不是拉丁语。这样做也是想方便读者辨识。古代的度量衡,如果能换算成英制,就加以换算。如没有必要,就用音译法加以处理。"②

"方便读者——以意补译。李善兰深知有的西方科学书籍,对中国读者来说都是一些新的知识。为了使读者更加容易明白译书中的知识内容,他有时并不拘泥于原文逐字逐句的直译,而是采用了'以意匡补'的翻译方式,即不仅采用非逐字逐句直译的'意译'方法,并且在意译译文之外他认为需要时附加了一些按语。许多按语都是对译文作进一步解说,这样就方便了不少缺乏相关专业知识的读者。"③

许渊冲在翻译《诗经》前言时总结:"……译诗如不传达原诗的音美,就

①② 《山海经》中华文库,前言介绍.

③ 方梦之,庄智象.中国翻译家研究[M].上海:上海教育出版社,2017.

不能保存原诗的意美。《诗经》总的说来是用韵的，译诗如不用韵，绝不可能产生和原诗相似的效果。恰恰相反，用韵的音美有时反而有助于传达原诗的意美。这就是说，用韵固然可能因声损义，不用韵则一定因声损义，用韵损义的程度反比不用韵小……我的英译希望尽可能传达《诗经》的意美、音美和形美。”①

由此可见，中国农业科技典籍翻译策略旨在平衡原文的忠实度与目标语言的可读性，采用音译保留专业术语，直译保持原文结构，意译确保内容流畅，注释辅助解释文化背景，增强理解，目标是让读者准确把握农业知识，同时体验原作风格等。

延伸阅读

Nongzheng Quanshu (also *Comprehensive Treatise on Agricultural Administration*) , was written by Xu Guangqi (1562—1633) in the late Ming Dynasty. Born in the Songjiang area in Jiangsu Province (now in Shanghai) where agriculture was very advanced, he was devoted to the agricultural development all his life. During the three-year period of mourning for his deceased father, he conducted in his hometown massive agricultural and ploughing experiments, and then went to Tianjing for similar experiments several times, the consequence of which was the completion of the agricultural encyclopedia. He died in his tenure of office, and it was his friend who helped publish the book in the 12th year of the reign of Emperor Chongzhen, i.e.in 1639. The book is divided into 60 volumes, including the agriculture-oriented thought, land policy, water conservancy, farm tools, arboriculture, sericulture, animal husbandry, relief polices and the like. Besides summarizing his experiences of succeeding in growing crops and cot-ton,

① 佚名.Book of Poetry 诗经[M].许渊冲，译.北京：五洲传播出版社，2011.

he went into details about water conservanolt and relief policies. He studied the lean years in the Chinese history, even delving into the concrete situations of the ill locust plagues which ever happened in China. The most effective way of putting an end to the lean years, he believed, was to establish the water conservancy projects on a large scale. In the Northwest, as he instantiated his point, instead of delivering grain from Southeast with great efforts, people should learn to reclaim the wasteland and establish the water conservancy projects on a large scale, turning Northwest into a major grain yielding area. A great number of reference books written before or at the time were recorded in *Nongzheng Quanshu*, and the understandings and perspectives of the author presented. It was a great agricultural work of special importance in the Ming Dynasty.

1. 词汇学习。

Ming Dynasty 明朝

agricultural encyclopedia 农业百科全书

land policy 土地政策

water conservancy 水利工程

farm tools 农具

arboriculture 园艺

sericulture 养蚕

animal husbandry 畜牧业

relief policies 救济政策

lean years 荒年

water conservancy projects 水利工程

wasteland 荒地

reference books 参考书籍

2. 句子翻译：请将以下材料翻译成汉语。

1) *Nongzheng Quanshu* (also *Comprehensive Treatise on Agricultural Administration*), was written by Xu Guangqi (1562–1633) in the late Ming Dynasty.

2) During the three-year period of mourning for his deceased father, he conducted in his hometown massive agricultural and ploughing experiments.

3) The book is divided into 60 volumes, including the agriculture-oriented thought, land policy, water conservancy, farm tools, arboriculture, sericulture, animal husbandry, relief policies and the like.

4) The most effective way of putting an end to the lean years, he believed, was to establish the water conservancy projects on a large scale.

5) A great number of reference books written before or at the time were recorded in *Nongzheng Quanshu*, and the understandings and perspectives of the author presented.

拓展知识

李约瑟

李约瑟(1900—1995)英国伦敦人，著名生物化学专家、汉学家，英国剑桥大学李约瑟研究所名誉所长。数次来到中国，先后任英国驻华科学参赞，中英科学合作馆馆长，1946年赴巴黎任联合国教科文组织自然科学部主任。

“早在1937年中日战争爆发时，他就是英国剑桥大学一位具有相当成就的生物学家，那一年，他不过三十多岁，前途无量，谁也想不到的是，一个偶然的机会，他在几位中国留学生的影响下，转而皈依中国文明，将毕生精力贡献给了中国科学技术史的研究事业。这几位中国留学生中有后来成为李约瑟太太的鲁桂珍，还有鲁桂珍的同学王应睐、沈诗章等人。这些成绩优异、聪明机智的中国留学生使李约瑟发现东方文明可能并不像西方主流学术界所说的那样毫无可取之处,而是在很多方面与西方近代文明比较接近,有许多共同的地方。所以,37岁的李约瑟决定学习中文,以便直接阅读中国典籍。

在著名汉学家夏伦教授指导下,李约瑟从阅读《管子》开始,边学汉语,边进行研究,通过几年探索,李约瑟确实迷上了中国文明尤其是中国人在科学技术和医学方面的成就。特别是魏特夫1931年发表的《为何中国没有产生自然科学》一文,对李约瑟产生了非常大的影响,直接激活了李约瑟研究中国文明史的兴趣,使李约瑟更热衷于探讨科学史上一些悬而未决的问题。

1942年,李约瑟受英国政府任命,前往中国担任英国驻华使馆科学参赞,稍后又受英国皇家学会委托,援助中国战时科学与教育机构,主要负责在重庆筹建中英科学合作馆,为中国科学界服务,包括提供文献、仪器、化学试剂以及传递科学信息、沟通中外科学界的联系等。李约瑟由此机会得以结识一大批中国一流科学家,这些学者有数学、物理、化学、工程、医学、天文、史学、考古、语言、经济、思想史、社会学等方面的专家。他们同李约瑟讨论了中国古代历史文化、科学发展和社会经济等一系列学术问题,很自然地提示李约瑟如欲研究中国文明应该读什么书、买什么书,并详细讲解每门学科史中的关键问题。在与中国学者的紧密交往中,李约瑟眼界大开,对中国文明的认识日趋加深,进一步坚定了研究中国文明的信心和决心,逐渐积累了足够数量的中国典籍,为其日后撰写《中国科学技术史》奠定了坚实基础。与很多中国学者不一样的是,李约瑟不仅致力于文献搜集、考订与研读,而且注重实地考察,在那短暂几年间,他在中国学者帮助下,实地考察了大半个中国,东到福建,西至敦煌千佛洞等文化遗迹,直接感受中国文明遗迹带来的震撼,获得大量研究灵感和启迪。”①

“李约瑟花费了半生的时间和精力,投入到《中国科学技术史》巨著的构思中。他说:中国的科学、技术和医学的历史将在学术史、经济史和社会史

① 马勇.青梅煮酒论英雄:马勇评近代史人物[M].南昌:江西人民出版社,2014.

上得到应有的地位，而且这将是一项永久持续的研究，就像所有的历史那样无限地继续下去。通过这部浩大恢宏的《中国科学技术史》，中国人记住了李约瑟这个响亮的名字；西方学者也了解到中国浩瀚的历史和文化曾深深地影响世界现代化的进程。李约瑟的《中国科学技术史》，勾勒了一幅完整表现中国古代科技发现和发明的画卷：火药、指南针、针灸、炼丹术……使西方读者第一次有可能较全面地认识中国对世界文化的贡献，也让许多中国人为祖宗留下来的宝贵财富而自豪。早在1964年，周恩来总理便指示，要促成《中国科学技术史》的翻译和出版，并由中国科学院中国自然科学史研究室（中国科学院自然科学史研究所的前身）负责具体组织这项工作。1975年，科学出版社出版了《总论》《天文》《数学》《地学》等共7分册的中译本，随即翻译和出版工作即告中断。1986年，李约瑟的这套巨著的翻译出版项目正式启动，国家对这项工作非常重视，由时任中国科学院院长卢嘉锡担任'李约瑟《中国科学技术史》的翻译出版委员会'主任委员，中国科学院自然科学史研究所专门成立了'李约瑟《中国科学技术史》翻译出版委员会办公室'，负责这部书的翻译和出版工作。从1990年开始，科学出版社与上海古籍出版社合作，再次出版这部专著。从1990年出版的第一卷《导论》，到2013年第六卷第六分册《生物学及相关技术：医学卷》，李约瑟《中国科学技术史》的翻译和出版工作经历了漫长的过程。"①

李约瑟作为最伟大的汉学家之一，创立了科技汉学新流派，打通了古今中西以及科学与人文的壁垒，堪称天下达人。回顾他的一生，你会感慨，东方文明和西方文明的最好融合其实是体现在人的身上。（《光明日报》评）

李约瑟对中国科技史的研究，改变了西方世界对中国文明落后的评价。（新华网评）

① 赵静荣.光明日报，2016-11-11.

第三章　翻译理论与农业科技典籍翻译

◎本章学习目标◎

1. 掌握翻译目的论的核心观点,并探讨其在科技典籍翻译中的应用
2. 理解生态哲学的基本理念,分析其在中国农业科技典籍翻译中的体现
3. 掌握古代饮食与酿酒文献中的术语翻译要点

第一节　翻译目的论与农业科技典籍翻译

翻译目的论(Skopos Theory)作为功能派的奠基理论,注重其翻译目的的选择意图,将"目的原则"作为翻译理论的最高准则,认为所有翻译活动应当遵循目的性、连贯性和忠实性三个法则[①]。"Skopos 这一术语通常用来指译文的目的,Skopos 一词来源于希腊语目的"Skopos"。德国翻译理论家汉斯·弗米尔(Hans J Vermeer)在此基础上还运用了"目标(aim)""目的(purpose)""意图(intention)"和"功能(function)"等词。为了避免概念混淆,诺德提议对意图和功能作基本的区分:"意图"是从发送者的角度定义的,而"功能"指文本功能,它是由接受者的期望、需求、已知知识和环境条件共同决定的。在弗米

① Munday J. Introducing Translation Studies[M]. London and New York: Routledge, 2016: 124.

尔的目的论框架中，决定翻译目的的最重要因素之一是受众——译文所意指的接受者，他们有自己的文化背景知识、对译文的期待以及交际需求。每一种翻译都指向一定的受众，因此翻译是在“目的语情景中为某种目的及目标受众而生产的语篇”。弗米尔认为原文只是为目标受众提供部分或全部信息的源泉。可见原文在目的论中的地位明显低于其在对等论中的地位。翻译目的论的核心概念是：翻译过程的最主要因素是整体翻译行为的目的”。农业科技典籍具有科技性与文学性，译者的翻译目的不同，采取的翻译策略也会有差异。

“翻译茶典籍的目的也是为了与其他民族进行平等的文化交流。从国人的理解角度将《茶经》再现给世界各国的读者……也可以保证对本土文化内涵更为精准的阐释，使中华文明对世界的贡献得以更广泛的传播与承认”（姜欣、姜怡，2009）。①

海外翻译家兼医学研究学者文树德（Paul U. Unschuld）在译本中提到了翻译《黄帝内经》的目的：他认为这些古代思想将有助于比较中西传统医学，帮助我们更好地理解医学的本质。只有通过这些翻译，我们才能追溯中国医学的发展历程，尤其是它如何在当代中国和国外重新定义为传统中医。他认为翻译对于促进中西医学传统之间的对话和交流至关重要。文树德的译文偏向海外医学研究的读者，因此在翻译时采取的策略会不同。

其观点可见：“If these ancient ideas are restored to life by our translation they will serve various useful purposes. First, these ideas will lend themselves to a comparison with similar traditions from the beginning of European medicine and may help us to gain a better understanding of ‘what is medicine.’”“For us to appreciate the basic differences and parallels between the more than millennia of

① 陆羽，陆廷灿.The Classic of Tea. The Sequel to The Classic of Tea Vol. Ⅰ，Ⅱ. 茶经.续茶经（全二册）[M].姜怡，姜欣，译.长沙：湖南人民出版社，2009.

Western and Chinese medical traditions, access to English translations of the seminal life science texts of Chinese antiquity, unadulterated by modern biomedical concepts and, is essential. Second, it is only on the basis of such translations that the later development of Chinese medicine can be traced, in particular its recent redefinition as Traditional Chinese Medicine in contemporary China and abroad. (Unschuld 9-10).”①

威斯(Ilza Veith)与李照国主要是为了推广中医文化,采取忠实于原文的翻译策略。同时,其他农业典籍的前言也可见译者翻译的目的,针对不同的读者群体采取了不同的翻译策略。如王宏与赵峥译者“这样做也是想方便读者辨识。古代的度量衡,各朝代标准不,如果能换算成英制,就加以换算。如没有必要,就用音译法加以处理。《梦溪笔谈》有个别篇幅非常难译,我们采取音译加注的办法处理,如有的译文中既出现英文,又出现汉字和其读音,这也是为了方便读者辨识。对《梦溪笔谈》中大部分所涉及的专有名词,我们大都采用了意译的方法,只有一些音律、象数中的名词等采用了音译法……总之,我们力求使自己的译文通顺、流畅和准确,并比原有的译文有所提高。我们希望这部《梦溪笔谈》英文全译本能为国内外更多的读者提供一个全面了解、研究《梦溪笔谈》的平台”②。

杨牧之学者在中华文库出版的英汉对照《天工开物》的前言写道:“西学仍在东渐,中学也将西传。各国人民的优秀文化正日益迅速地为中国文化所汲取,而无论西方和东方,也都需要从中国文化中汲取养分。正是基于这一认识,我们组织出版汉英对照版《大中华文库》,全面系统地翻译介绍中国传统文化典籍。我们试图通过《大中华文库》,向全世界展示,中华民族五千

① Unschuld, P.U., translator.Huang Di Nei Jing Su Wen:Nature, Knowledge, Imagery in an Ancient Chinese Medical Text.Edited by Bing Wang.U of California P, 2003.

② 沈括.梦溪笔谈全译[M].胡道静,金良年,胡小静,译.上海:上海古籍出版社,2008.

年的追求，五千年的梦想，正在新的历史时期重放光芒。中国人民就像火后的凤凰，万众一心，迎接新世纪文明的太阳。”①可见其翻译目的也是为了传播中华典籍文化。

在翻译中，译者也会采取一些特殊的方式对译文进行处理，以便读者更深入理解原文的意思。例如，在《黄帝内经》的译文前言中，译者李照国对括号的使用做了说明如下：“1. 圆括号()：置于有关音译术语或概念之后，所括内容为该术语或概念的现行译法或解释，如 Jing (Essence), Shen (Spirit or Mind)等。2. 方括号[]：译文将因行文或表达之需所增加的词语置于其中，以明确何为译文，何为解释，如在 If it is inserted too deep, [it will] cause intermal damage 一句中，中括号内的词语均为翻译时因行文和表达之需所增加的内容，故置于中括号之中。3. 大括号{}：由于年代久远辗转传抄，《黄帝内经》里时有衍文出现。对于这些衍文，翻译时均置于{}之中。”②

【双语赏析】

《黄帝内经》之《脏气法时论》篇

肝色青，宜食甘，粳米、牛肉、枣、葵皆甘。心色赤，宜食酸，小豆、犬肉、李、韭皆酸。肺色白，宜食苦，麦、羊肉、杏、薤皆苦。脾色黄，宜食咸，大豆、豕肉、栗、藿皆咸。肾色黑，宜食辛，黄黍、鸡肉、桃、葱皆辛。辛散，酸收，甘缓，苦坚、咸耎。

毒药攻邪、五谷为养，五果为助，五畜为益，五菜为充。气味合而服之，以补精益气。此五者，有辛、酸、甘、苦、咸，各有所利、或散，或收，或缓，或急，或坚，或耎，四时五脏，病随五味所宜也。

---

① Tian Gong Kai Wu 天工开物[M]. 王义静，王海燕，刘迎春，等译. 广州：广东教育出版社，2011.

② 闫畅，王银泉. 中国农业典籍英译研究：现状、问题与对策(2009-2018)[J]. 燕山大学学报(哲学社会科学版)，2019，20(3)：49-58.

Zangqi Fashi Lunpian: Discussion on the Association of the Zang-Qi with the Four Seasons

The liver is related to blue [in colors] and sweet flavor is good for the liver. Polished round-grained nonglutinous rice, beef, Chinese date and sunflower are all sweet [in taste]. The heart is related to red [in colors] and sour flavor is good for the heart. Red bean, dog meat, plum and chives are all sour [in taste]. The lung is related to white [in colors] and bitter flavor is good for the lung. Wheat, mutton, apricot and macrostem onion are all bitter [in taste]. The spleen is related to yellow [in colors] and salty flavor is good for the spleen. Soy bean, pork, chestnut and the leaves of beans are all salty [in taste]. The kidney is related to black [in colors] and pungent flavor is good for the kidney. Yellow millet, chicken, peach and scallion are all pun-gent [in taste]. Pungent [flavor] disperses, sour [flavor] as tringes, sweet [flavor] relaxes, bitter [flavor] hardens and salty [flavor] softens.

Duyao (drugs) [can be used] to attack Xie (Evil), the five kinds of grain[1] [can be used] to nourish [the body], the five kinds of fruit[2] [can be used] to assist [the five kinds of grain to nourish the body], the five kinds of domestic animals[3] [can be used] to supplement [the Five Zang-Organs] and the five kinds of vegetables[4] [can be used] to enrich [the viscera]. Harmonic mixture of proper flavors can supplement Jing (Essence) and nourish Qi. These five kinds of food have pungent, sour, sweet, bitter and salty [flavors] respectively and tonify [certain Zang-organs or Fu-organs] by means of dispersion or astringency, moderation or promptitude[5], hardening or softening. [In treating diseases,] these five flavors should be used in accordance with [the changes of] the four seasons and [the states of] the Five Zang-Organs.

Notes

[1] Wang Bing (王冰) said, "[The five kinds of grain] include Jingmi (粳米, polished round-grained nonglutinous rice), Xiaodou (小豆, red bean), Mai (麦, wheat), Dadou (大豆, soy bean) and Shu (黍, broomcorn millet).

[2] Wang Bing (王冰) said, "[The five kinds of fruit] include Tao (桃, peach), Li (李, plum), Xing (杏,apricot), Li (栗, chestnut)and Zao (枣, date or jujube).

[3] Wang Bing (王冰) said, "[The five kinds of domestic animals] include Niu (牛, cow or ox), Yang (羊, goat or sheep), Shi (豕, pig) Quan (犬, dog) and Ji (鸡, chicken or rooster).

[4] Wang Bing (王冰) said, "[The five kinds of vegetables] includeKui (葵, sunflower), Huo (leaves of beans), Xie (薤, acrostem onion), Cong (葱, scallion) and Jiu (韭, Chinese chives).

[5] The original Chinese for "promptitude" is Ji (急) which literally means "hurry" or "urgency". According to Suwenshi (《素问识》), Ji (急) perhaps is a redundancy due to misprinting or miscopying. (李照国译)①

第二节 生态哲学与农业科技典籍翻译

在农业科技典籍中,融合了生态文化哲学观的理念,强调了人与自然的和谐共生。通过尊重自然规律、保护生态环境、倡导可持续发展等方式,促进农业生产方式向更加生态友好的方向转变。这种生态文化哲学观的拓展不仅

① 闫畅,王银泉.中国农业典籍英译研究:现状、问题与对策(2009—2018)[J].燕山大学学报(哲学社会科学版),2019,20(3):49-58.

是对传统农业科技的延续和升华,更是对人类与自然和谐共生的追求和实践。

《天工开物》主要根植于中国的固有文化传统。天工开物取自“天工人其代之”及“开物成务”,体现了朴素唯物主义自然观,与当时占正统地位的理学相异。这种异端化的思想趋势,反映着一种新的社会现象和时代取向。但是,个人的思想可以有异于主流,却不能超脱于时代。古代素以农业作为重中之重,所以宋应星的文章中也处处体现出贵五谷轻金玉的思想。[①]

《农政全书》提倡农业生态观与中国农业智慧,书中指出:“旱田获刈才毕,随即耕治晒暴,加粪壅培,而种豆麦蔬茹,因而熟土壤而肥沃之,以省来岁功役。”连续种植会导致土壤肥力下降,因此建议交替种植不同的作物,并定期让土地休耕以恢复其自然肥力。这些生态农业观点代表了中国传统农业的哲学智慧。书中指出根据土地的肥瘠和面积来确定播种量,这一原则实际上是为了防止过度开发和浪费,注重资源的合理使用和有效分配,体现了哲学里的“和谐”生态农业文化。《农政全书》强调了优质土壤对农业生产的重要性,提倡对土壤进行保育和改良以增加肥力,如适量施用农家肥以改良土壤,这正是生态农业中尊重和保护自然资源的理念。在此书中,阐述的鸡粪可以作肥,养蚕可以利用桑树叶,桑树又可以提供阴凉等,都体现出农业生态系统的多元和平衡,体现了生态农业里“万物循环,相互依存”的中国哲学智慧。

《齐民要术》认为,在农业生产中必须以“天人合一”的和谐观为基础,以“顺自然”与“骆马首”相结合的生态观为指引,做到顺应天时、迎合地利与循物之性,并重视人的主体能动作用,在农业耕种中应依循天、地、人、物四者的共生、共促与共享。

从《要术》体现的伦理主旨来看,深受传统农业社会顺时而化、顺势而行的生态伦理思想影响,其阐述的社会生产与生活境况都是遵循着“道法自

① 宋应星:《天工开物》.人民网,2015-12-23.

然”与“天人合一”的朴素生态农学观，其从保持人的主体性、能动性和独立性出发，将构成农业生产的人、物与天地资源浑然于一体，体现了对自然规律的敬畏，并推崇人与自然之间的共促、共建与共享的和谐生态。这种充分平衡农业、环境、人力三者之间关系的生态农学观是对自然规律的基本遵循，与人类尊重自然、敬畏自然、融入自然、与自然共享、“天人合一”的逻辑主旨具有内在一致性，更与现代生态伦理所倡导的人与自然和谐共生理念构成价值的趋同性。①

《茶经》的生态自然思想可从茶的实物到器皿，再到水的选择，以及各地风俗的呈现，茶的华夏版图也变得清晰可见，到最后形成的是茶的图腾与仪式，《茶经》所要表达的意图也十分明了：人要把自己的精神融合在格物运化之中，只有与自然浑为一体，才能再回到自然。②

《农政全书》较为全面地反映了徐光启以“农”“政”辩证关系为基础，展现了经济、技术与农业生产部门相统一的“大农业”系统观和生态观。《农政全书》的精要之处也在于：徐光启并没有仅仅将农业问题拘泥于对以往农业科学知识的总结，而是将目光放到了更为长远的政治生态上，将农政措施和农业技术相结合，使《农政全书》超越了以往的纯技术性农业书籍，集中表达了徐光启以农治国的农业生态观。③

《吕氏春秋》多次强调农业的可持续发展与生态环境的保护，这种农业生态思想体现着可持续发展的生态伦理观，如《吕氏春秋·圜道》体现了自然万物循环往复的法则，强调了农业生产应顺应自然规律，如季节变化，以实现农作物的生长和收获。《吕氏春秋》强调了天、地、人的统一关系，认为人的行为和治理应该顺应天地的自然规律。“上揆之天，下验之地，中审之人”，意

① 潘才宝，刘湘溶.《齐民要术》的生态伦理取向探究[J].长春理工大学学报（社会科学版），2018，31(4)：57-60.

② 纯道.禅艺茶道[M].上海：文汇出版社，2017.

③ 韩忠治.《农政全书》中的朴素农业生态观[N].学习时报，2022-06-28.

味着人类应该观察和学习自然界的运行规律,并将其应用于人类社会的治理和个人行为中。这种思想倡导人们模仿自然界的和谐与平衡,以达到社会和环境的可持续发展。《吕氏春秋》被认为是具有浓郁稷下道家底色的生态哲学,它综汇了古代思想,紬绎出"贵生之道,安宁之道,听言之道"三大纲领,这些思想都与生态哲学密切相关,强调了人与自然的和谐相处。《吕氏春秋》通过"十二纪"的编排,体现了对自然规律的尊重,如"春生,夏长,秋收,冬藏"的自然规律,来指导人类社会制度和人事运行的法则。这种思想体现了人类行为应与自然规律相符合,以达到生态的平衡和稳定。

同时,生态哲学观与文化之间存在着密切的关系。生态哲学观强调人类与自然的和谐共生,提倡尊重自然、保护环境、促进可持续发展。这种观念不仅影响着人们对自然和生态系统的态度,也深刻影响着人们的文化观念和行为方式。生态哲学观认为对自然的尊重和崇敬,对生态平衡和生物多样性的重视,以及对可持续发展和环保意识的强调。这些价值观和观念在文化中得到传承和弘扬,影响着人们的生活方式、价值取向和社会行为。文化强国的建设已经成为新时代的文化使命,农业典籍的挖掘与对外传播助力于文化强国建设,为我国农业文化传承与海外传播提供有力保障,探究农业典籍中的农业生态文化哲学智慧路径有助于凝聚我国民族文化,传承中国传统油脂文化,树立中国形象,对外传播中国哲学文化智慧。文化自信、开放包容、守正创新以及为国家建设和民族复兴注入精神力量的重要性。作为新时代的人民,我们既要自信地传承和发展中华民族的优秀传统文化,同时也要积极吸收借鉴世界各国的优秀文化成果。我们要通过文化建设来增强国家的软实力,凝聚全国人民的力量,涵养全民族的精神,为实现中华民族的伟大复兴作出贡献。

【双语赏析】

"何以说天道之圜也？精气一上一下，圜周复杂，无所稽留，故曰天道圜。何以说地道之方也？万物殊类殊形，皆有分职，不能相为，故曰地道方。"

——《吕氏春秋·圜道》

译文一：Why is the principle of Heaven round? Yin is ascending and Yang is descending, they travel ceaselessly in circles, so the principle of Heaven is round. Why is the principle of Earth square? Everything of the world is different both in shape and in function. Therefore it cannot be replaced by anything else. So, the principle of Earth is square.

—The Round Principles 翟江月 译①

译文二：Why do we say that the way of heaven is circular? The essence of life and vital energy move one above the other in repeated circles and never come to a stop. That is why we say the way of heaven is circular. Why do we say that the way of earth is square? The multitude of things are different in category and shape and each of them has its own role to play, which cannot be exchanged. That is why we say that the way of earth is square.

—The Circular Way 汤博文 译②

"凡十二纪者，所以纪治乱存亡也，所以知寿夭吉凶也。上揆之天，下验之地，中审之人，若此则是非可不可，无所遁矣。"

——《吕氏春秋·序意》

---

① 吕不韦. The Spring and Autumn of Lü Buwei Vol. Ⅰ, Ⅱ, Ⅲ. 吕氏春秋(全三册)[M]. 翟江月，译. 桂林：广西师范大学出版社，2005.

② 吕不韦. Lü's Commentaries of History 吕氏春秋[M]. 汤博文，译. 北京：外文出版社，2010.

译文一：These twelve records are aimed at recording the significant events in history, such as order, disorder, survival or perdition of the states, and explaining the reason for occurrences in human life, such as longevity, mortality, fortunes or mishaps. We are doing that by observing the will of Heaven above, examining occurrences on Earth and inspecting human affairs in between, so all mistakes in judgement, true or false, right or wrong, will be avoided this way.

—The Postscript of the Twelve Records①

译文二：The twelve groups of Records comment on peaceful and riotous times and on the survival and demise of states, from which longevity or premature death, auspiciousness or misfortune may be deduced. When it is measured against heaven above, tested against earth below, and examined by men in the middle, there will be no misjudgment of right or wrong and of what is permissible or what is not permissible.

—Postscript②

第三节 古代饮食习俗特点及其翻译技巧

古代饮食习俗在中国传统文化中扮演着重要的角色,展现出丰富多彩的特点。古人对饮食的重视可从其独特的审美情趣和文化内涵中得以体现。在古代,人们在不同的节日和场合会有特定的饮食习俗,如春节吃年糕、端午节吃粽子、中秋节品尝月饼等。这些食物不仅是为了满足口腹之欲,更是表达节日祝福和纪念传统文化的重要方式。古代饮食习俗的丰富

① 宋应星.《天工开物》.人民网,2015-12-23.

② 潘才宝,刘湘溶.《齐民要术》的生态伦理取向探究[J].长春理工大学学报(社会科学版),2018,31(4):57-60.

多样，反映了古人对饮食的独特理解和对传统文化的珍视。

首先，要深入了解古代饮食习俗背后的文化内涵和历史背景，以便准确理解和翻译相关内容。其次，在翻译过程中要尽量保持译文的原汁原味，避免过度解释或删改，以保留原文的特色和风格。此外，选择恰当的词语和表达方式也是至关重要的，使译文通顺自然，同时准确传达古代饮食习俗的意义。最后，针对一些特定的古代饮食习俗，适当进行文化解释或注释，有助于帮助读者更好地理解和欣赏古代文化。通过运用这些翻译技巧，翻译者可以更好地传达古代饮食习俗的特点和文化内涵，促进不同文化之间的交流与理解。

【双语赏析】

《吕氏春秋》第14卷《本味篇》

（以下译文一选自译者翟江月，译文二选自译者汤博文）

汤得伊尹，祓（音弗）之于庙，爝（音决）以爟（音灌），衅以牺猳。明日设朝而见之，说汤以至味。

汤曰："可对而为乎？"对曰："君之国小，不足以具之，为天子然后可具。夫三群之虫，水居者腥，肉玃（音觉）者臊，草食者膻。恶臭犹美，皆有所以。凡味之本，水最为始。五味三材，九沸九变，火为之纪。时疾时徐，灭腥去臊除膻，必以其胜，无失其理。调合之事，必以甘、酸、苦、辛、咸。先后多少，其齐甚微，皆有自起。鼎中之变，精妙微纤，口弗能言，志不能喻。若射御之微，阴阳之化，四时之数。故久而不弊，熟而不烂，甘而不哝，酸而不酷，咸而不减，辛而不烈，淡而不薄，肥而不腻。"

译文一：After Tang acquired Yi Yin, he held a ceremony at the central ancestral temple for his sake. During the ritual, bundles of reeds were lit up to drive away demons for him and boars were killed for a blood sacrifice. The next day, Tang held court to interview Yi Yin and Yi Yin described the most delicious food for Tang.

"Would you make these delicacies for me?" asked Tang. "Sorry," replied Yi Yin, "Your territory is too small to prepare these things. You can only enjoy them after you unify the world and become a Son of Heaven yourself. Of these three kinds of animals, the aquatic animals are fishy, the carnivorous animals are stinking and the herbivorous animals smell of mutton. As for these creatures, whether they are odoriferous or look malformed, they are all of some use. Blue beards and licorice are useful as well. In the concoction of flavour, the crucial factor is water. When boiling the five kinds of seasonings and the three kinds of ingredients many times to make the flavour change correspondingly, the crucial factor is fire. In order to remove those aforementioned unpleasant smells, fire is indispensable. But make sure that the duration and degree of heating is suitable. In order to concoct flavour, it is always necessary to use sweet, sour, bitter, pungent or salty seasonings and gradients. Nevertheless, in order to concoct various kinds of flavour, these things should be added in the correct sequence and the right amount should be used since flavour can be affected by subtle changes in this process. Changes of flavour taking place inside the cauldron are as subtle as the techniques of archery or horsemanship, the cooperation of Yin and Yang and the order of the four seasons, so they cannot be explained with words, nor can they be sensed by insight. So, even though the food might have been cooked for a long time, it will still be good. It is well done but not overcooked. It is not too sweet, too sour, too salty, too pungent, too insipid or too fat1y.

译文二：After having enlisted the service of Yin Yi, Tang held an exorcist ceremony in the ancestral temple, lighted reed torches to dispel bad luck and smeared the sacrificial vessels with the blood of a pure-colored boar. At court the following morning, he received Yin Yi with decorum, and Yin Yi spoke to him about delicacies.

"Can these delicacies be obtained and prepared?" asked Tang. The answer was: "Your state is too small and does not have them. You can have them when you become the king. There are three kinds of creatures. Those living in the water have a fishy smell; those eating flesh, a foul smell; those munching grass, a mutton-like smell. Unpleasant smells can be transformed into nice smells. The way lies in how to prepare them. Water is the most important thing in preparing the different favors. When preparing the five flavors, using the three basic materials, bringing the pot to the boil repeatedly and changing the favor repeatedly, the most important thing is regulating the heat, which can be very hot sometimes and gentle at others. The fishy, foul and mutton-like smells must be removed by heat, which must be appropriately handled. Food must be seasoned with sweet, sour, bitter, hot or salt flavors. The time for adding the flavors and the amount to be added, which is always very small, are regulated by rules. The change in the favor of the food in the tripod is so subtle that it cannot be expressed in words but can only be perceived by intuition like the subtleties in the skills of shooting an arrow or driving a chariot, the merging the Yin and Yang essences and the change of the four seasons. Therefore, the food is cooked for a long time without destroying it, done without overheating it, sweet but not excessively so, sour but not excessively so, salty without losing its original favor, hot but not too hot, light but not tasteless, succulent but not greasy.

肉之美者：猩猩之唇，獾獾之炙，隽觾之翠，述荡之腕，旄象之约，流沙之西，丹山之南，有凤之丸，沃民所食。

译文一：Lips of gorillas, feet of badgers, tails of swallows, hooves of the sacred two-head horses and waists of yaks are the best flesh in the world. Moreover,phoenix eggs produced at the place west of Liu Sha and south of Dan Mountain are delicacies enjoyed by the people of the state of Wo.

译文二：Among the delicious meat are monkey's lips, wild boar's trotters, the tail meat of cuckoos. the small legs of Shudang, the tails of yaks and elephants, and phoenix eggs produced south of Danshan Mountain and eaten by people of the state of Wo.

鱼之美者:洞庭之鱄,东海之鲕。醴水之鱼,名曰朱鳖,六足,有珠百碧。雚水之鱼,名曰鳐,其状若鲤而有翼,常从西海夜飞,游于东海。

译文一：The cowfishes of the Dong Ting Lake, the roes of the East Sea, the red turtles with six feet and green-pearl-like scales of the Li River, the carp-like skates with wings of the Guan River, which often fly from the West Sea to the East Sea are the most delicious fishes in the world.

译文二：Among the delicious fish are porpoise from Lake Dongting, Er from the East Sea, a kingfisher-colored fish named Zhubie with six feet and emitting bubbles, and a fish named ray which is like the carp in the River Guanshui but with wings which often fly from the West Sea to the East Sea at night.

菜之美者,昆仑之苹,寿木之华,指姑之东。中容之国,有赤木玄木之叶焉,馀瞀之南,南极之崖,有菜,其名曰嘉树,其色若碧,阳华之芸,云梦之芹,具区之菁。浸渊之草,名曰土英。

译文一：A kind of duckweed produced by Kun Lun Mountain, the flowers of the ever-living tree, leaves of the red and black trees produced in the state of

Zhong Rong east of Zhi Gu, the emerald-like vegetable called Jia Shu growing on the cliffs of the South Pole south of Yu Mao, the coleworts of the Yang Hua Lake, the celeries of the Yun Meng Lake, the leeks of the Ju Qu Lake and a weed named Tu Ying of the Jin Yuan Lake are the best vegetables.

译文二：Among the delicious vegetables are duckweed from the Kunlun Mountains, Huaguo from Shoumu, the leaves of Red Tree and Black Tree from the state of Zhongrong east of Zhigu, a jade-colored vegetable called Jiashu from the southern Land's End south of Yumao, rue from Lake Yanghua, water celery from Lake Yunmeng, rape turnip from Lake Juqu and a grass named Tuying from Qinyuan.

和之美者：阳朴之姜，招摇之桂，越骆之菌，鳣鲔之醢，大夏之盐，宰揭之露，其色如玉，长泽之卵。

译文一：The gingers of Yang Pu, the cinnamon of Zhao Yao, the bamboo shoots of Yue Luo, the minced flesh made of sturgeons, the salt of Da Xia, the emerald dew of ZaiJie and the birds' eggs of Chang Ze are perfect seasonings.

译文二：Among the delicious condiments are ginger from Yangpu, cinnamon from Zhaoyao, bamboo shoots from Tuoyue, sturgeon and tuna paste, salt from Daxia, a sauce from Zaijie with the color of a white jade, and bird's eggs from the Great Lake.

饭之美者：玄山之禾，不周之粟，阳山之穄，南海之秬。水之美者：三危之露，昆仑之井，沮江之丘，名曰摇水，曰山之水，高泉之山，其上有涌泉焉，冀州之原。

译文一：The rice of Xuan Mountain, the millet of Bu Zhou Mountain, the yellow millet of Yang Shan and the black millet of the South Sea are the best flavoured grains. The dew drops of San Wei Mountain, the spring of Kun Lun Mountain, the spring flowing across the hills along the bank of the Ju River called the Yao Water, the water of Yue Mountain, the spring of Gao Quan Mountain—which is also the head spring of the waters of Ji Zhou are the best waters.

译文二：Among the delicious cereals are rice from Xuanshan, millet from Mount Buzhou, glutinous millet from Yangshan and black corn from the South Sea. Among delicious water are dew drops from the Sanwei Mountains, spring water from the Kunlun Mountains, a spring water called Yaoshui from the hills by the Zujiang River, water from White Mountain and water from the gushing springs on Gaoquan Mountain which are the headwaters of the rivers in Jizhou.

果之美者：沙棠之实，常山之北，投渊之上，有百果焉，群帝所食；箕山之东，青鸟之所，有甘栌焉，江浦之橘，云梦之柚。汉上石耳。

译文一：The fruit of the crab apple trees of Kun Lun Mountain, the fruit enjoyed by God and other deities produced at the place north of Chang Mountain above Tou Yuan, the grapefruit produced at the place east of Ji Mountain where the Green Bird lives, the mandarin orange produced along the riverbanks of the Yangtze River, the shaddocks of the Yun Meng Lake and the stone ear (a kind of vegetable) are the best fruits.

译文二：Among the delicious fruits are the fruit of crabapple trees, fruits from north of Changshan Mountain above Touyuan Pool enjoyed by the early kings, sweet hawthorn from east of Mount Qishan where the bluebirds live, orange from the banks of the Changjiang, pomelo from the banks of the Yunmeng River and mushrooms from either side of the Hanshui River.

所以致之，马之美者：青龙之匹，遗风之乘。非先为天子，不可得而具。天子不可强为，必先知道。道者止彼在己，己成而天子成，天子成则至味具。故审近所以知远也，成己所以成人也。圣人之道要矣，岂越越多业哉！"

译文一：Anyone who wants to taste all these wonderful things must use the best horses—named 'the Green Dragon' or 'the Flying Wind' to transport them. And it is not possible to obtain these things unless you can unify the whole world and become a Son of Heaven yourself. However, the Son of Heaven cannot be enthroned by force. In order to become a Son of Heaven, you should master Tao first. And in order to master Tao, you should always resort to yourself instead of resorting to others. If you can master Tao, you can become a Son of Heaven. lf you can become a Son of Heaven, you can enjoy all these delicacies. Hence, you can know about things taking place in remote areas by examining those occurring nearby, and you can edify others if you can master Tao yourself. Nevertheless, since sages' ways of doing things are very simple and concise, is it really necessary for them to take a lot of concrete actions in person?"

译文二：To ship these delicacies it requires the finest of horses, such as the Green Dragon and the Racing Wind. Before one becomes the king, one cannot enjoy these delicacies. One cannot become the king by arbitrary decision. Before one can become the king, one must understand the way of virtue and justice. The way of virtue and justice is not to be found in others but within oneself. When one has mastered the way of virtue and justice, one will naturally become the king. When one has become the king, there will be delicacies to be enjoyed. Therefore, looking closely into what is close at hand enables one to understand what is far away. Only when one has become accomplished can one help others. The way of the sages is simple. Need one busy oneself with so many things?"

第四节　古代醋食技术制作翻译技巧

醋的本义,《说文》谓"客酌主人也"。《玉篇》谓"进酒于客曰献,客答主人曰醋,"是客人答谢主人的礼仪。可见。醋之古今意义相去甚远。醋在古代的称谓有醯、酢、苦酒等名。春秋战国时期称之为醯,《论语·公冶长》:"子曰:熟谓微生高直?或乞醯焉,乞诸其邻而与之。"疏:"醯,醋也。"《论语》是说,有人向微生高讨醯,他家没有,于是到邻家讨些再给那人,可见公元前六世纪,醋已经是生活中常用的东西了。宋史绳祖《学斋占毕卷四·九经所无之字》谓:"九经中无醋字,止有醯和用酸而已,至汉方有此字。"说明醋字在汉代方出现。汉以前称醋为醯或用酸。①

醋的制作与酒密不可分,因酒放置日久氧化而变化成醋,所以古人又称醋为酢酒、淳酢、酢浆、苦酒等。苦酒作为醋名,当在东汉魏晋时期。东汉医学家张仲景在《伤寒论》中云:"少阴病,咽中伤,生疮,不能语言,声不出者,苦酒汤主之。"在《金匮要略》中治疗"以汗出入水中浴,水从汗孔入得之"的黄汗病时,亦用有苦酒。北魏贾思勰在《齐民要术·作酢法》所列作苦酒法数种,皆指醋。陶弘景谓"酢酒为用,无所不入,愈久愈良。以有苦味,俗呼苦酒"。葛洪《肘后方》亦用苦酒疗疾治病。可见。苦酒作为醋名在魏晋东汉之时已为常见。到了唐代,醋、酢、苦酒之名相混而用,亦无明确之区分。《千金要方》:"治小儿重舌方,赤小豆末,醋和涂舌上";"治胞衣不出方。鸡子一枚,苦酒一合,和饮之";"治舌卒肿,满口溢出……半夏十二枚,洗熟,以酢一升煮,取八合。稍稍含漱之。"醋作为中药的一种,生活常用之调味品,与"客酌主人"之义相去远矣"。

① 张胜忠.古代醋名小识[J].吉林中医药,1992(2):46-47.

关于醋的制作方法，历代多载于农学书和本草学书上。虽然醋的种类很多，但医用则仅以米醋入药。据《本草纲目》载，米醋的制法乃“三伏时，用仓米一斗，淘净，蒸饭，摊冷，罨黄，晒簸，水淋净……和匀入瓮，以水淹过，密封暖处，三七日成矣”。明代徐光启《农政全书》也载有“作酢法”。醋、酢之名在唐代以后渐趋一致，沿用至今。[①]

【双语赏析】

醋[②]

释名：酢音醋。醯音兮。苦酒〔弘景曰〕醋酒为用，无所不入，愈久愈良，亦谓之醯。以有苦味，俗呼苦酒，丹家又加余物，谓为华池左味。〔时珍曰〕刘熙《释名》云：醋，措也。能措置食毒也。古方多用酢字也。

集解：〔恭曰〕醋有数种：有米醋、麦醋、曲醋、糠醋、糟醋、饧醋、桃醋，葡萄、大枣、蘡薁等诸杂果醋，会意者亦极酸烈。惟米醋二三年者入药。余止可啖，不可入药也。

〔诜曰〕北人多为糟醋，江外人多为米醋，小麦醋不及。糟醋为多妨忌也。大麦醋良。

〔藏器曰〕苏言葡萄、大枣诸果堪作醋，缘渠是荆楚人，土地俭啬，果败则以酿酒也。糟醋犹不入药，况于果乎？

〔时珍曰〕米醋：三伏时用仓米一斗，淘净蒸饭，摊冷盦黄，晒簸，水淋净。别以仓米二斗蒸饭，和匀入瓮，以水淹过，密封暖处，三七

① 许明武，王烟朦．中国科技典籍英译研究（1997—2016）：成绩、问题与建议[J]．中国外语，2017，14（2）：96-103.

② 李时珍．Condensed Compendium of Materia Medica. Vol. Ⅰ，Ⅱ，Ⅲ，Ⅳ，Ⅴ．本草纲目选（全五册）[M]．罗希文，译．北京：外文出版社，2012.

日成矣。糯米醋:秋社日,用糯米一斗淘蒸,用六月六日造成小麦大炮和匀,用水二斗,入瓮封酿,三七日成矣。粟米醋:用陈粟米一斗,淘浸七日,再蒸淘熟,入瓮密封,日夕搅之,七日成矣。小麦醋:用小麦水浸三日,蒸熟盦黄,入瓮水淹,七七日成矣。大麦醋:用大麦米一斗,水浸蒸饭,盦黄晒干,水淋过,再以麦饭二斗和匀,入水封闭,三七日成矣。饧醋:用饧一斤,水三升煎化,入白麹末二两,瓶封晒成。其余糟、糠等醋,皆不入药,不能尽纪也。

CU

Vinegar

—Drug of inferior class *Mingyi Bielu (Records of Famous Doctors)*.

[Explanation of Names]

XI KU JIU

Tao Hongjing: When vinegar and wine are used as drugs, they function well in the body. The longer they are stored, the better the quality. Vinegar is also known as Xi. As it is bitter in taste, it is also called Kujiu (meaning"bitter wine"). Taoist alchemists blend vinegar with other drugs and use it as a seasoning in their processing.

Li Shizhen: The book *Shi Ming* by Liu Xi—Vinegar is something that can detoxify food toxins. In ancient times, the character used was 酢 (Cu).

[Previous Explanations]

Su Gong: Vinegar can be varied in accordance with the substances used: rice vinegar, wheat vinegar, vinegar made of Jiuqu, Mikang/spermo-dermis oryzae/seed-coat of rice, Jiuzao/vestigium vinum/remains of wine-brewing, Yitang/saccharum granorum/maltose, peach, grape and date. Some of them, if well, brewed, are very sour and strong. But only rice vinegar that has been stored for two to three years can be used in medicine. Vinegar made of other things can only be used as seasoning.

Meng Xian: People in the north brew vinegar with Jiuzao. People living south of the Yangtze River brew vinegar with rice. Wheat vinegar is not very good in quality. Vinegar made of Jiuzao has a lot of taboos in its use. Vinegar made of barley is of good quality.

Chen Cangqi: Su Gong said that grape and date can be brewed to make vinegar. This is because Su was a native of the Jing and Chu area (middle Yangtze valley). That was a place where life was harsh. So, even rotten fruits were used to brew wine. Even vinegar made of Jiuzao cannot be used in drugs, not to mention that brewed from fruits.

Li Shizhen: The brewing of vinegar may vary as follows:

Rice vinegar: In Sanfu days (three 10-day periods of the hottest days in the year), rinse one *dou* of long-stored rice and then steam it. Spread it to cool it down and seal it until a yellow coating appears. Dry the drug in the sun and winnow away the chaff. Rinse it in water again. Steam another two *dou* of long-stored rice and blend it with the previous processed lot. Keep it in a jar and immerse it in water. Seal the jar tightly and keep in a warm place. Vinegar is made after 21 days.

Vinegar made of Nuomi: On Qiu she day [the fifth Wu day after the day of the solar term marking the Beginning of Autumn (August 7), about mid-September], blend the following things evenly: One *dou* of Nuomi, rinsed and steamed, and Jiuqu made of Xiaomai, brewed on the sixth day of the sixth month. Keep them in a jar and add in two *dou* of water. Seal the jar. After 21days, vinegar is brewed.

Vinegar made of Sumi (millet): Rinse one *dou* of long-stored Sumi and soak for seven days. Steam it and rinse it. Keep the thing in a jar and seal it. Stir it in the daytime and the evening. Vinegar will be brewed after seven days.

Vinegar made of Xiaomai (wheat): Soak Xiaomai in water for three days. Steam it until it is well done. Seal it until a yellow coating appears. Keep it in a jar immersed in water. Vinegar will be brewed after 49 days.

Vinegar made of Damai (barley): Soak one *dou* of Damai in water and steam it until it is well done. Seal it in a jar until a yellow coating appears. Rinse with water. Blend another two *dou* of steamed Damai with the previous processed lot. Keep it in a jar immersed in water. Seal it. Vinegar will be ready after 21 days.

Vinegar made of Yitang (maltose). Stew one *jin* of Yitang in three *sheng* of water until it melts. Add in two *liang* of white Jiuqu. Seal it in a bottle and keep it in the sun.

Vinegar made of Jiuzao and Mikang are not good for medical use.Brewing of them and other vinegars are not recorded here.

(罗希文 译)

第五节　古代酿酒技艺的文化翻译技巧

中国古代的制曲与酿酒技艺十分精湛。了解农业科技典籍中的酿酒技术，传承中国传统酿酒技艺，有利于传播中国文化，推进中国酿酒技术走向世界。我国独特的制曲技术是古代劳动人民一大发明。中国酒曲酿酒的发明在6—7世纪（大约隋唐时）由朝鲜传入日本和越南等国家，形成了用曲酿酒的东方酒文化的特色。从《本草纲目》中所说的"烧酒非古法也"可见烧酒在元代以前是没有的。"关于酒的起源传说，在古籍中如《吕氏春秋·勿躬篇》《战国策》《黄帝内经·素问·汤液醪醴论》《淮南子·说林训》《酒诰》等都有详细记载。"①

"《齐民要术》中对影响发酵过程中的水、温度和加料等因素作了详细记述，其中对制曲、酿酒用水的认识和创造对水采取的"间歇灭菌"法；发酵醪液中二氧化碳气体和酶的早期认识，发酵过程中对微生物散热措施的应用等记载，有些是世界酿酒史料巾最早或最先进的记录。"②

"用酒曲酿造是中国酒的特色，是我们祖先的伟大创造。我国传统酒有黄酒和白酒，都是用不同酒曲酿造而成的。酒曲品种繁多，有大曲、小曲、麦曲、红曲等，各有不同的功效。而酿酒工艺各有特点，如大曲白酒工艺采用大曲（有酱香型、浓香型、清香型、兼香型等）为发酵剂进行固体窖内发酵、固态蒸馏的方法；如小曲白酒采用小曲进行固态或固液结合酿造；而黄酒生产工艺采用小曲、麦曲或红曲为糖化发酵剂，进行先固态、后液态发酵而成。

---

① 傅金泉．中国古代科学家对酿酒发酵化学的重大贡献[J]. 酿酒，2012，39(5)：87-91.

② 杨勇．试论《齐民要术》中的我国古代制曲、酿酒发酵技术[J]. 西北农林科技大学学报(自然科学版)，1985(4)：55-64.

酿制工艺内容丰富多彩,酒品繁多,酒质各有特色,代代相传。[①]

"《齐民要术》中,虽然神曲的糖化发酵能力最强,笨曲的糖化发酵能力最弱,但在用于酿酒时,却是笨曲使用广。"[②]"使用神曲的酿造方法主要有三斛麦曲法、造酒法(包括作林与黍米酒法、糯米酒法)、神曲梗米醪法、神曲酒方法(渍曲法)、河东神曲造酒法(浸曲法)等;使用白醪曲做酒仅仅一种方法,家酿白醪法;而使用笨曲造酒的方法多达20种,包括春酒法(春酒曲,浸曲法)、浸药酒法(加草药发酵)、颐酒法(颐曲,浸曲法)、河东颐白酒法、笨曲桑落酒法、笨曲白醪酒法(浸曲法)、蜀人作酴酒法、粟米酒法、粟米炉酒法、魏武帝上九温法、《食经》作白醪酒法、《食经》冬米明酒法、《食经》夏米明酒法、朗陵何公夏封清酒法、愈赌酒法、勐酒法、夏鸡鸣酒法、柯椎酒法等。还有一种酿酒方法,是二种曲同时使用的,特别是神曲与笨曲同时用于发酵,如粱米酒法、穄米酎法和黍米酎法。不加曲的浸泡药酒二种,即和酒法(草药浸泡酒)、《博物志》胡椒酒法(加草药加热浸泡)。"

有关酿酒的文字记载,最早出现在《周礼》中。《周礼》是记录周代礼制最为详备的著作,又称《周官》,记载先秦时期社会政治、经济、文化、礼法等,涉及内容非常丰富,堪称"中国文化史之宝库"。"《本草纲目》中,李时珍对酒的系统分类,为后来酒的食用、医疗、药用价值方面提供了较科学的应用标准,这其中,有关药酒的记载和阐述有着巨大的研究价值。药酒不仅是对酒的酿造文化的发展和创新,更对当代酒的常饮和入药都有积极的借鉴意义。《天工开物》中提到用明矾水(无机物溶液)培养纯化红曲种(微生物),这种方法沿用至今。"[③]《北山酒经》是我国现存的第一部关于酿酒工艺的专著,是宋代制曲酿酒工艺理论的代表作。

---

① 刘欣.探析古代酿酒技术[J].黑龙江史志,2015(5):295.

② 范文来.《齐民要术》中的中国古代酿酒技术[J].酿酒,2020,47(6):111-113.

③ 张萌,刘俊仙.丝绸之路与古代中国蚕桑技术的外传[J].中国民族博览,2015(8):184-186.

古籍中的酿酒技术

古籍名称	特点
《天工开物》卷下《曲蘖》	从制作工艺方面用绘图方式讲述了酒母、神曲和丹曲的加工步骤。
《齐民要术》卷七有《造神曲并酒》篇、《白醪酒》篇、《笨曲并酒》篇和《法酒》篇	唐朝以前最系统、最完整的一部古代酿酒著述，操作性和指导性都很强。
《周礼》	用“五齐三酒”对酒进行分类：“辨五齐之名，一曰泛齐，二曰醴齐，三曰盎齐，四曰缇齐，五曰沉齐。辨三酒之物，一曰事酒，二曰昔酒，三曰清酒。”
《礼记·月令》	比较完整的酿酒工艺流程：“仲冬之月，乃命大酋，秫稻必齐，曲蘖必时，湛炽必洁，水泉必香，陶器必良，火齐必得。兼用六物，大酋兼之，毋有差贷。”
《周礼·天官》	周朝设置了酿酒的官职记载：“酒正，中士四人，下士八人，府二人，史八人，胥八人，徒八十人。酒人，奄十人，女酒三十人，奚三百人。”
《北山酒经》	宋代酿酒专著，上卷论酒的发展历史，中卷论制曲，下卷记造酒。“酸米人甑，蒸汽上，用杏仁五两(去皮尖)、蒲萄二斤半(浴过干、去子皮)，与杏仁同于砂盆内一处，用熟浆三斗，逐旋研尽为度，以生绢滤过，其三斗熟浆，泼饭软盖，良久出饭摊于案上，依常法候温，入曲搜拌。”
《本草纲目》	“烧酒非古法也，自元时始创其法。用浓酒和糟入甑，蒸令气上，用器承取滴露。凡酸坏之酒，皆可蒸烧。近时，惟恐以糯米或粳米或黍或大麦蒸熟，和曲酿瓮中七日，以甑蒸取。其清如水，味极浓烈，盖酒露也。”
《史记·大宛列传》	“宛左右以蒲桃为酒，富人藏酒至万余石，久者数十年不败。”张骞从大宛引进葡萄的同时，还引进了酿酒技术。

【双语赏析】

酿酒:沈括《梦溪笔谈 汉人酿酒》①

原文:汉人有饮酒一石不乱,予以制酒法较之,每粗米二斛,酿成酒六斛六斗,今酒之至醨者,每秫一斛,不过成酒一斛五斗,若如汉法,则粗有酒气而已,能饮者饮多不乱,宜无足怪。然汉一斛,亦是今之二斗七升,人之腹中,亦何容置二斗七升水邪?或谓"石"乃"钧石"之"石",百二十斤。以今秤计之,当三十二斤,亦今之三斗酒也。于定国饮酒数石不乱,疑无此理。

今释:汉代人有喝1石酒不醉的,我拿酿酒法一比较,汉代每2斛粗米,酿成6斛6斗酒;现在最薄的酒,每一斛稻谷不过酿成一斛五斗酒。假如像汉代的方法,就略有酒气罢了,能喝的人多喝而不醉,应该不值得奇怪。但汉代的一斛,也就是现在的二斗七升,人的肚子里又怎么装得下二斗七升呢?有人说"石"是"钧石"的"石",是120斤。用今天的秤来称,应当是32斤,也就是现在的三斗酒。于定国喝几石酒不醉,我怀疑没有这种道理。

译文: It was said that in the Han Dynasty there were people who could drink one *dan* of rice wine without getting drunk.* I investigated into their way of making rice wine and found that at that time two *hu* of unpolished rice could produce six *hu* and six *dou* of wine.* Currently one *hu* of rice can only produce one and a half *hu* of the lightest wine. So the wine made in the Han Dynasty could be only called "the drink that contains little alcohol." It was quite usual for heavy drinkers to drink a lot of wine made in this way and still remain sober. However,one *hu* in the Han Dynasty equals two *dou* and seven *sheng* today. How

① 沈括.梦溪笔谈全译[M].胡道静,金良年,胡小静,译.上海:上海古籍出版社,2008.

could a man's belly hold two *dou* and seven *sheng* of wine? Some people argue that as the unit of weight, one *dan* equals 120 jin in the Han Dynasty. According to the present weighing system, one *dan* equals thirty-two *jin* or three *dou* of wine. When l read the sentence "Yu Ding guo drinks several *dan* of wine without getting drunk" in *The Book of Han*, I could hardly believe it.

Translator's Notes

*When used as the unit of capacity. "石" was pronounced as "*shi*" in ancient China and "*dan*" in Modern Chinese. One dan is equal to 10 *dou* or 100 liters.

*In ancient China, *hu* was also widely used as the unit of capacity. One *hu* is equal to one *dan*.

（王宏，赵峥 译）

练　习

1. 句子翻译。

1）羹之有菜者用梜，其无菜者不用梜。（《礼记·曲礼》）

2）御同于长者，虽贰不辞，偶坐不辞。（《礼记·曲礼》）

3）东风解冻，蛰虫始振，鱼上冰，獭祭鱼，鸿雁来。（《礼记·月令》）

4）始雨水，桃始华，仓庚鸣，鹰化为鸠。（《礼记·月令》）

5）蝼蝈鸣，蚯蚓出，王瓜生，苦菜秀。（《礼记·月令》）

6）《农政全书》："六经中无茶，荼即茶也。《毛诗》云：'谁谓荼苦，其甘如荠'，以其苦而味甘也。"（《续茶经》陆廷灿）

7）"夫茶灵草也，种之则利溥，饮之则神清。上而王公贵人之所尚，下而小夫贱隶之所不可阙，诚民生食用之所资、国家课利之一助也。"（《续茶经》陆廷灿）

8)唯独介子推拒绝接受封赏。他带母亲隐居绵山,不肯出来。晋文公无计可施,只好放火烧山,逼其下山。谁知介子推母子宁愿被火烧死也不肯出来。为了纪念介子推,晋文公下令将绵山改名叫介山,并修庙立碑。同时,还下令在介子推遇难的这一天,“寒食禁火”,举国上下不许烧火煮食,只能吃干粮和冷食。

9)寒食节是在清明节的前一天,古人常把寒食节的活动延续到清明,久而久之,清明取代了寒食节。拜介子推的习俗也变成了清明扫墓祭拜祖先。为了使纪念祖先的仪式更有意义,应该让年轻一代的家庭成员了解先人过去的奋斗历史。

2.请用翻译目的论翻译以下内容。

煮醴酪制作技巧

准备材料:将牛奶、醴粉、白糖、水等材料准备好。

煮牛奶:将牛奶倒入锅中,加入适量的水,用中火煮沸。

加入醴粉:将醴粉加入锅中,不断搅拌,直到醴粉完全溶解。

加入白糖:将白糖加入锅中,继续搅拌,直到白糖完全溶解。

煮醴酪:将锅中的牛奶继续煮沸,不断搅拌,直到醴酪变得浓稠。

倒入容器:将煮好的醴酪倒入容器中,放置冷却。

提示:翻译中需要注意使用准确的词汇和语言表达,以确保翻译的准确性和通顺性。需要注意文化差异,如“醴粉”在英语中可以翻译为sweet potato starch。翻译中需要注意使用正确的语态和时态,以确保翻译的准确性和通顺性。

3.辨析译文:你认为哪个译文更加体现了翻译的目的论呢?

1)本草名鼠尾草,一名陵翘。出黔州及所在平泽有之,今钧州新郑岗野间亦有之。苗高一二尺叶似菊花叶,微小而肥厚;又似野艾蒿叶而脆,色淡绿。(《救荒本草》)

译文一：Shuju[鼠菊, European verbena herb, Verbena officinalis L.], also named Shuweicao, Qing and Lingqiao, grows in the weedy wetlands of Qianzhou. Presently, it can be seen in the hills and fields of Junzhou and Xinzheng. Its plants are 1–2 Chi high. Its green leaves are like the chrysanthemum leaves, smaller and thicker, and also like the wild mugwort leaves, more fragile. ①

（范延妮 译）

译文二：In the herbal records, it is known as Sage, also called Lingqiao. It grows in the plains of Qianzhou and other similar wetlands, and now it can also be found in the wilds of Xinzheng among the hills of Junzhou. The plant grows to a height of one or two feet, with leaves resembling those of chrysanthemums, but smaller and more succulent; they also resemble the leaves of wild wormwood, yet are more brittle, with a light green color.

2)湿种之期，最早者春分以前，名为社种（遇天寒有冻死不生者），最迟者后于清明。凡播种，先以稻、麦稿包浸数日，俟其生芽，撒于田中，生出寸许，其名曰秧。秧生三十日即拔起分栽。

译文一：At the earliest, the soaking of seed rice can be done before the Spring Equinox which is known as "she planting" (If there is a cold weather during this period of planting, the seeds will be frozen and can't grow out of the ground.), or can be done after Pure Brightness at the latest. The seed rice is wrapped in rice or wheat straw and soaked for a few days.②

（王义静、王海燕、刘迎春 译）

译文二：The optimal timing for sowing seeds is before the Spring Equinox, a period known as "she zhong" (though there's a risk of seeds freezing and not

---

① 朱橚.《救荒本草》汉英对照[M].范延妮，译.苏州：苏州大学出版社，2019.

② Tian Gong Kai Wu 天工开物[M].王义静，王海燕，刘迎春，等译.广州：广东教育出版社，2011.

germinating due to cold weather). The latest it should be done is after the Qing-ming Festival. For sowing, first, wrap the rice and wheat straw in cloth and soak it for several days until sprouts emerge. Then, scatter these sprouted seeds in the fields. When the seedlings grow to about an inch tall, they are called "saplings". After thirty days of growth, the saplings are then uprooted and transplanted.

3)升明之年,正阳之气主治,其德普及四方,五行气化平衡。其气上升,其性急速,其作用为燃烧,其生化为繁荣茂盛,其属类为火,其脏心,心其畏寒,其主舌,其谷麦,其果杏,其实络,其应夏,其虫羽,其畜马,其色赤,其养血,其病瞤瘛,其味苦,其音徵,其物脉,其数七。(《黄帝内经》)

译文一:

"In the year of Shengming (elevation and brightness), the Upright Yang is in predomination. [Under such a condition,] [Fire] influences everywhere; the transformation of the five [kinds of Qi] is in balance; its Qi elevates; its properties are swift; its function is burning;and its transformation is prosperity. [In terms of correspondence, it corresponds to] fire in categorization;shining in administration; hotness in weather; heat in manifestation; and the heart in the viscera. The heart detests cold and governs the tongue. [It corresponds to]wheat in crops; apricot in [the category of] fruits; threadlike structure [in the parts of] fruits; summer in seasons;winged ones in [the category of] insects; horse in animals;red in colors; blood in its tonification; tremor, spasm and convulsion of muscles in diseases; bitterness in tastes; Zhi in scales; pulse in constituents; and seven in numbers."

(李照国 译)

译文二: In the year of ascending brightness, the righteous and positive energy presides over the natural world, spreading its grace to all directions, with the transformations of the Five Elements in balance. This energy rises upward, characterized by its swift and intense nature, acting as combustion, fostering the flourishing growth of all things. It corresponds to the element of fire, is associated with the heart, which fears cold, primarily affects the tongue, with wheat as its corresponding grain, apricots as its fruit, the seed's outer layer known as the "network", aligns with the summer season, corresponds to feathered creatures among animals, horses among livestock, the color red, aids in blood circulation, may lead to conditions like skin blemishes or moles, has a bitter taste, corresponds to the "zhi" note in music, and the number seven.

4)多生山阴近水处。数根丛生,一根数茎,茎大如箸,其涎滑。其叶两两对生,如狗脊之叶而无锯齿,青黄色,面深背浅。其根曲而有尖嘴,黑须丛族,亦似狗脊根而大,状如伏鸱。(《本草纲目》)

译文一:It grows on the shady side of mountains, close to water. It grows in tussocks, each with several roots, Its root is the size of a chopstick. When broken, it secretes slippery juice. Its leaves grow in opposition in pairs. It looks like Gouji/folium cibotii/leaf of East Asian tree fern, but without the serrated edge. It is blue-green-yellow, darker on the face and lighter at the back, Its root is curled, with a sharp end, and it has fine black hair-like roots thickly attached, It is similar to Gouji/rhizoma cibotii/rhizome of East Asian tree fern, but somewhat bigger. It looks like the head of a sparrow hawk.①

译文二:It grows abundantly in the shaded areas near water on mountains. Several roots cluster together, with each root having multiple stems, the size of which is as thick as chopsticks, and they are slippery with a slimy substance. The leaves grow in pairs opposite each other, similar to the leaves of a dog's back but

① 李时珍.Condensed Compendium of Materia Medica. Vol. Ⅰ, Ⅱ, Ⅲ, Ⅳ, Ⅴ.本草纲目选(全五册)[M].罗希文,译.北京:外文出版社,2012.

without serrated edges, with a greenish-yellow color, darker on the top and lighter on the back. The roots are curved with a pointed tip, surrounded by a cluster of black hairs, resembling the roots of a dog's back but larger, and they resemble a crouching owl in shape.

延伸阅读

The highly technical nature of the text makes it exceptionally difficult to read, especially since the early Chinese had no systematic method for identifying plants. Names tended to differ from area to area and period to period. Even those names that attained general acceptance frequently lacked standard characters, thereby resulting in the unsystematic use of phonetic loans. Moreover, throughout history the meaning of even some well established names underwent a process of change, thus creating even more ambiguity. To make matters worse, later scribes tended to be ignorant of such subjects as soil types and plant life so that the names and technical terms involved were especially prone to corruption. In fact, were it not for the painstaking work of numerous Chinese scholars, especially Xia Weiying's efforts at identifying the names of plants and trees, this translation would have been impossible.

It is also unfortunate that while much of the chapter's material is clearly based on real conditions and practical observation, its objectivity is sometimes marred by the attempts of its author or authors to adhere to the formulaic demands of Five Phases correlative ideology. This is particularly true of the first and last sections of the text where artificial categories such as the five colors and correlates such as the five notes are assigned to each category of soil. It is also difficult to tell just what the primary purpose of this chapter may have been.

Unlike most chapters of the Guanzi, including the previous chapter (XVIII, 57), which deals with irrigation and water control, "Di yuan" provides no instruction on how its factual content may be used for political or other practical purposes. Be that as it may, the work has much to tell us about Chinese understanding of their environment, and especially their understanding of the need to relate agricultural production to soil types.

CONTENT AND STRUCTURE

The text may be divided into five major sections, the first of which describes, in unrhymed prose, five categories of alluvial or irrigated soil, their characteristic plants and trees, and the depth of their water tables, ranging from seven to thirty-five Chinese feet. It also lists a correlate musical note in the Chinese pentatonic scale for each soil type, a color for its water, and the special characteristics of its people. Attached to the end of this section is a rather lengthy explanation of the five notes, what they sound like, and the formula for determining the length of the tubes used in the standard pitch pipes. It appears likely that this explanation, partially written in rhyme, is a later insertion and not part of the original text.

The second section, also in unrhymed prose, gives the names of fifteen types of hill land classified according to altitude and depth of the water table. No information concerning such things as characteristic plants and trees is provided for the soils listed in this section.

The third section, written entirely in rhyme, discusses five different levels of mountain terrain, beginning at the highest level at which normal plant life exists and moving downward. For each level it provides the distance down to the water table and the names of characteristic plants and trees.

The fourth section, unrhymed, is very short but of special interest in that it consists of a list of twelve plants considered characteristic of a twelve-level hydrological sequence, beginning with water plants such as the lotus and water chestnut and moving up the scale through various dry land plants to a floss grass variety of sedge.

The fifth and final section ranks the soils of the nine regions of ancient China into three grades with six subdivisions in each.The first three sub-divisions of the highest grade are treated in considerable detail, including an extensive discussion, all in rhyme, of soil characteristics and typical plants and trees. The remaining subdivisions are treated very briefly in unrhymed, formulaic statements limited to listing the chief characteristics of each soil type, its characteristic grains, and its level of productivity in relation to the first three subdivisions described at the beginning of the section.①

1.根据以上内容,运用批判性思维选出最佳答案。

1) How should the phrase "unsystematic use of phonetic loans" be accurately translated into Chinese based on the context provided in the paragraph?

A. 音韵借词的无计划使用

B. 音素借词的随意使用

C. 音标借词的不系统化使用

D. 音位借词的无规律使用

2) In the given text, what translation technique would be most suitable for conveying the concept of "formulaic demands of Five Phases correlative ideology" into Chinese?

① 摘自W-Allyn Rickett在《管子》(地员)第58章的部分前言内容。W-Allyn Rickett. GUANZI—Political, Economic, and Philosophical Essays from Early China[M]. Princeton: Princeton University Press, 1985.

A. 五行对应意识形态的公式性要求

B. 五行对应意识形态的常规性需求

C. 五行对应意识形态的固定要求

D. 五行对应意识形态的刻板要求

3) How could the sentence "It also lists a correlate musical note in the Chinese pentatonic scale for each soil type, a color for its water, and the special characteristics of its people." be best translated into Chinese to maintain the original meaning?

A. 它还为每种土壤类型列出了相应的中国五声音阶中的音名,其水的颜色,以及该地区人民的独特特征。

B. 它还列出了每种土壤对应的中国五声音阶的音符,其水的颜色,及其人民的特别属性。

C. 它还为每种土壤类型指定了中国五声音阶中的音乐符号,其水的颜色,和那里的人的特殊性格。

D. 它还为每种土壤类型提供了中国五声音阶中对应的音符,其水的颜色,以及那里人们的特殊特征。

4) Translate the sentence "Unlike most chapters of the Guanzi, 'Di yuan' provides no instruction on how its factual content may be used for political or other practical purposes" into Chinese.

A. 与《管子》的大多数篇章不同,《地员》未提供如何将其事实内容用于政治或其他实际目的的指导。

B. 与《管子》大多数章节不同,《地员》未提供如何将其事实内容用于政治或其他实际目的的指导。

C. 与《管子》大多数章节不同,《地员》没有说明其事实内容如何用于政治或其他实际目的。

D. 与《管子》大多数章节不同,《地员》未提供其事实内容如何用于政治或其他实际目的。

5) How would you translate the phrase "the first three sub-divisions of the highest grade are treated in considerable detail" into Chinese?

A. 对最高等级的前三个子类别进行了详细探讨。

B. 最高等级的前三个子分部受到了详细处理。

C. 最高等级的前三个子分部被详细描述。

D. 最高等级的前三个子分部被详细论述。

6) What is the title of the first section of the original text?

A. Translation Techniques

B. Linguistic Difficulties

C. Cultural Interpretation

D. Problems of Translation

2. 根据原文,思考以下问题。

Discuss the significance of the author's inclusion of the five categories of alluvial or irrigated soil in the first section of the text. How does this detailed description contribute to the overall understanding of the relationship between soil types and agricultural production in ancient China?

拓展知识

徐光启

徐光启(1562—1633),字子先,号玄扈,上海人,生于明嘉靖四十一年(1562年),卒于崇祯六年(1633年)。中国明末杰出的科学家、思想进步家、翻译家和爱国政治家。他最早将翻译的范围从宗教、文学扩大到了自然科学。徐光启是中西文化交流的先驱之一,是上海地区最早的天主教基督徒,作为热忱而忠贞的教友领袖和护教士，被誉为明代“圣教三柱石”之首（圣教三柱石是指明朝时天主教耶稣会传教士利玛窦在中国传教其间所训练出的第一代基督徒里最有成就的三个人，他们是徐光启、李之藻与杨廷筠）。东学西渐的主要代表人物之一，是中国翻译西方科学文献、介绍西方近代科学的先驱。编撰了《农政全书》《崇祯历书》《徐氏庖言》《测量法义》《勾股义》等农学、天文、军事著作，还与利玛窦合译了《几何原本》，与熊三拔合译了《泰西水法》，与龙华民等人合译了《崇祯历法》。

第四章　农业科技典籍中的植物典籍翻译

◎本章学习目标◎

1. 了解植物典籍定义和特点
2. 掌握植物典籍术语翻译技巧
3. 能够对植物典籍双语文本进行基本赏析

第一节　农业科技植物典籍的概述

植物典籍属于农业科技典籍中的一种，也称为本草学。"本草"一词首见于《汉书·郊祀志》"方士使者副佐、本草待诏七十余人皆归家"，后多用于药典名称，如《救荒本草》《神农本草经》与《本草纲目》等。本教材所指的植物典籍指的是所有典籍中包含了植物的典籍翻译。农业典籍中包含一些中药材植物典籍，一些古籍中包含的植物类翻译和一些专门性学科类植物典籍。主要有《神农本草经》《救荒本草》《本草纲目》《植物学》《黄帝内经》和《茶经》等。其中包含大量的农业词语，如《救荒本草》《本草纲目》等，植物典籍的语言朴实，涵盖内容兼具药用与食用方面。《救荒本草》成书于明代，明永乐四年(公元1406年)刊刻于开封,是我国历史上第一部以救荒为宗旨的植物学著作。是周定王朱橚就藩于开封之时组织人员撰绘的一部植物图谱,为我国史上首部以救荒为宗旨的农学、植物学专著,是一部专讲地方性植物并结合食用方面以救荒为主的植物志。全书分上、下两卷,记载植物414种,每种都配有精美的木刻插图,其中出自历代本草的有138种,新增276种,从分类上分为:草类245种、木类80种、米谷类20种、果类23种、菜类46种,按部编目。《救荒本草》兼具农业与医药典籍特征,书中含有大量的农业词语,主要包括农业名物词和植物描述词。其中以农业名物词数量最多,仅植物名称词就有一千余个。《救荒本草》受到世界植物学家和科学史家的高度赞赏。

"《植物学》一书是由英国传教士韦廉臣(Alexander Williamson,1829—1890)、艾约瑟(Joseph Edkins,1823—1905)和中国学者李善兰(1811—1882)合作编译,于清咸丰八年(1858年)由上海墨海书馆出版。这是中国第一本介绍西方近代植物学基础知识的著作。全书共八卷,约35 000字,插图200多幅,根据英国著名植物学家J.林德利(John Lindley,1799—1865)1848年出版的《植物学基础》(*The Elements of Botany*)等书的相关内容编译而成。韦廉臣和艾约瑟都是非常热心传播生物学知识的传教士,后者还是一位汉学家。《植物学》的内容包括植物的广泛用途以及植物学的重要性,植物在地球上不同纬度的分布,植物的形态和根、干、枝、叶、花、果实、种子等器官形态构造和生理功能,以及由林德利所创的自然分类系统。书中将植物分为303科,简介了其中37科,还简单介绍了植物学观察和实验的方法。李善兰较好地处理了汉译植物学过程中存在的新名词和术语的问题。全书创译了不少当代植物学名词和术语,如花瓣、萼、子房、心皮、胎座、胚、胚乳和细胞,分类等级的'科',以及若干科的名称,如伞形科、石榴科、菊科、唇形科、蔷薇科、豆科等,将'botany'翻称'植物学'一词也是他们创造的。除植物学内容外,书中还介绍了一些自然神学方面的内容。自《植物学》传入日本后,书中所用的'植物学''科'等名词术语也被日本科学界所采纳,沿用至今。《植物学》虽未能作为教会学校之教科书,但它对中国生物学教育的发展起到了重要的作用。对一些上层社会的知识分子接触西方科学知识、开阔他们的视野、进而致力于科学技术的引进,也发挥过一定的启蒙作用。"[①]

"《植物学》一书并不是专门针对某一部英文植物学专著的全译本,而是受到译者主观能动性的驱使及晚清客观社会环境的制约而进行的带有明确科学启蒙目的的选择性译本。学界目前普遍认为,《植物学》所据之外文原本为英国植物学家林德利(John Lindley,1799—1865)的植物学著作。"[②]

《植物学启蒙》成书于1886年,为艾约瑟(Joseph Edkins)所编译的《格致启蒙十六种》中的重要组成部分之一。与《植物学》及《植物图说》相比,《植物学启蒙》中除了涉及西方植物学的基础知识,也论及了植物学的教授方法等内容。《植物学启蒙》原作者为胡克(J.D.Hooker),该书分为30章,涉及植物的结构、营养、分类、植物学教授法等内容,并附有数十幅图[③]。后经考证,《植物学启蒙》中所附图的数量为68幅[④]。该书前25章共163节介绍植物的结构、营养和分类,第26章是植物学实验,第27至29章则是植物学教授法内容,最后的第30章相当于一个索引,回顾前面曾经提到的各种植物名录。

① 中国大百科全书.

② 汪子春.我国传播近代植物学知识的第一部译著《植物学》[J].自然科学史研究,1984(1):90-96.

③ 付雷.晚清中下学生物教科书出版机构举隅[J].科普研究,2014(6):61-72.

④ 孙雁冰.晚清(1840—1912)来华传教士植物学译著及其植物学术语研究[J].山东科技大学学报(社会科学版),2019,21(6):33-38.

第二节　农业科技植物典籍术语的翻译技巧

“农学典籍记载了我国先进的传统农业文明成果，跨越时间长，数量大且涉及门类众多，在世界农业科技史上地位居高。大多数农业典籍术语产生自古代，因此受到古代传统哲学思想的影响，具有自身独特性”……总结其具有历史性、人文性与民族性。①

《植物学》翻译中“不用确定术语而是通过具体描述来表达新的概念”，如“年轮”的概念用“外长类每岁多一层，断木验其层数，能知木生之年数也”来表达。②

根据张翮学者的观点，术语的翻译主要有以下方法：第一，《植物学》术语翻译拒绝采用音译。音译是翻译外国地名和人名的常用方法，但很少用于科学翻译，除了有机物名称的翻译。李善兰等人在翻译植物学术语时，几乎杜绝了音译的使用，仅有的“淡巴箍科”使用了音译（“淡巴苏”是烟草“tobacco”的音译）。这样做不仅便于理解，也提升了术语传承的可能性。据研究，19 世纪通过音译方法翻译成中文的科学名词术语，几乎全部被取代了，这很可能与中文和英语等外文的语言差距大而产生的音译借词障碍有关。值得一提的是，《植物学》中的某些外国人名和地名，以及一些动植物名称，采用了音译，较有代表性的有“礼乃亚（Linnaeus，林奈）巴巴西（Pampas，潘帕斯）”“阿低泥亚（actinia，海葵）”“白和白（baobab，猴面包树）”等。还有一些词语的翻译一半是音译，另一半却属意译，如“维纳斯

① 袁慧，冯炜．基于目的顺应论的农学典籍《齐民要术》科技术语翻译研究[J]．长春理工大学学报（社会科学版），2023，36(6)：158-162.

② 罗桂环．我国早期的两本植物学译著——《植物学》和《植物图说》及其术语[J]．自然科学史研究，1987，6(4)：383-387.

之蝇牢(Venus's fly-trap,捕蝇草)""阿萝萝番薯(arrowroot,竹芋)",这类情况比较少见。①

第二,"一词多译"和"多词一译"现象比较常见。因此,一个外文术语被译成多个中文术语,或者多个含义不同的外文术语被译为同一个中文术语的情况在《植物学》中并不少见。前一种情况如上文提到的"cel1"被译成"细胞""胞体""小室""子房(室)"等多个词,其复数"cells"还偶尔被译为"聚胞体";再如"ovule"除了被译为"(小)卵",也偶尔被译为"心""ovary"除了被译为"子房",还被译为"心之本"。后一种情况如"medullary sheath(髓鞘)"和"carpel(心皮)"都被译为"心皮""ovule(胚珠)""pistil(雌蕊)""pith(髓)"都被译为"心"。还有些意思相近但还是存在区别的几个词,都被译为同一个术语,如"albumen""perisperm""nutritive matter""nourishing matter"。

第三,术语依据原文文本的范围,不超过原文文本的术语范围。

第四,特别注重增加译词的本土化色彩。晚清科学译著大多具有本土化色彩,特别是早期的译著。而这一特征在《植物学》的术语翻译上有较为突出的体现,许多译词都带有浓重的本土化色彩,体现出与中国传统文化的密切联系。如将"double flower"译为"重台"这个传统文化中标示"复瓣花"的词语,既恰当地表达了意思,又承载了本土文化的因子。

除此以外,其他学者也提出了植物学术语翻译相关技巧:

李宇明学者认为,"术语民族化,即术语本土化,是指将外语中的科技术语引入本民族语言中,其实质是改变术语的语言形式,用本族语言翻译外国科技术语。"②周有光学者认为,"因此西学术语汉译时,术语民族化提倡尽量用意译的翻译方法,使术语切合本土语言的使用情况,创造有本土化特色的

① 张翮.基于"双语对校"的晚清译著《植物学》研究[D].合肥:中国科学技术大学,2017.

② 李宇明.谈术语本土化、规范化与国际化[J].中国科技术语,2007(4):5-10.

名词。”[①]“《植物学》中植物学术语在音节上符合经济性原则，简单明了的双音节词语和以双音节为基础的多音节词语，具有明显的民族化特征。”[②]

“李善兰在介绍近代西方植物学基础知识时，尽量运用我国传统植物学中已有的名词术语，采用直观、浅显、通俗的表达方式，以便人们联系已有的知识更好地理解和接受新的知识。关于这点还突出地反映在他对植物科名的翻译上。他译科名主要是根据这几个途径：一是根据我国有关这科的传统的植物类群的集合名词，如豆科、瓜科、五谷科；二是把为人们所熟知的该科典型植物作为科名，如芭蕉科、菱科、莲科等大部分科都如此，对后来的命名影响较大；三是根据该科植物花的形态来翻译的，如伞形科、十字科、唇形科等，其‘伞形’‘唇形’等都是袭自我国传统的本草书中。”[③]

“第一，对农学植物术语的翻译，因民族间的文化共性，译者多采用直译的翻译策略，使读者与作者间对话能够顺利进行，提高了译文阅读的流畅性，实现了译者的翻译目的。第二，对农学技术术语的翻译，译者主要采用直译与意译相结合的翻译策略，使自我文化在他者文化中保留了自己独有的色彩，保留下来的原文文化要素给英文读者以充分的想象空间。第三，对农学作物术语的翻译，译者采用了音译加注释的翻译策略，虽然增加了原文术语的描写信息，但降低了英文读者的阅读障碍，顺应了读者的需求。”[④]

“《齐民要术》中的典籍科技术语与中华民族的思维方式、审美方式、哲

① 周有光. 漫谈科技术语的民族化与国际化[J]. 中国科技术语，2010(1)：8-10.

② 周倩儒. 论《植物学》术语译名的民族化[J]. 新楚文化，2022(10)：61-64.

③ 罗桂环. 我国早期的两本植物学译著——《植物学》和《植物图说》及其术语[J]. 自然科学史研究，1987(4)：383-387.

④ 袁慧，冯炜. 基于目的顺应论的农学典籍《齐民要术》科技术语翻译研究[J]. 长春理工大学学报(社会科学版)，2023，36(6)：158-162.

理思想紧密相连,具有概念符号和文化内涵的混合特质,体现着语言、社会、历史和建构的意义。”[①]如石声汉的《齐民要术》中的植物术语翻译:

原文:“粟、黍、穄、粱、秫,常岁岁别收:选好穗纯色者,劁才雕反刈,高悬之。”

译文:Seed corns for spiked millet, ordinary and glutinuous panicled millets, ordinary and glutinous Setaria, are always to be collected separately every year. Pick out plump ears uniform in colour ; cut them down and hang up to dry[②]. 译者采用音译加注释的方法翻译了无法对等的植物术语。

除此之外,在一些具有药用性和食用性的植物典籍也比较常见。“19世纪以来,西方学者对《救荒本草》产生了浓厚兴趣,开始对书中所载植物进行分析和整理。1881年,德国植物学家布赖特施耐德(E.BretSchneider)在伦教出版《中国植物志》一书,对《救荒本草》中的176种植物的学名进行了考证鉴定,认为书中优秀的木刻图比欧洲要早70年。”[③]

对于新术语的翻译,译者在翻译说明中写道:“《救荒本草》中部分中药名称经多方考证,至今仍无法确定为何种中药,在翻译成英文的时候,采取音译加medicinal的方法;翻译成拉丁文的时候,采取materiamedica加音译的方法。”[④]在《神农本草经》中,神农向人们传授了许多农业技术和知识,如农耕、种植、养蚕等。据传说,神农还曾亲自下田教人耕种,并发明了许多农具和器具,如犁、耙、镰刀等,使中国农业生产得到了显著的提高。翻译时为了

---

① 王烟朦.基于《天工开物》的中国古代文化类科技术语英译方法探究[J].中国翻译,2022,43(2):156-163.

② A Preliminary Survey of the Book Ch'i Min Yao Shu 2nd ed. 齐民要术概论(第二版)[M].石声汉,译.北京:科学出版社,1962.

③ 张帆.安徽大农业史述要[M].合肥:中国科学技术大学出版社,2011.

④ 朱橚.《救荒本草》汉英对照[M].范延妮,译.苏州:苏州大学出版社,2019.

体现英文的形合，汉英转换时候可灵活变译。原文中为否定，翻译时采取反译法，将“不”译为肯定的语气。如“久服，补髓益气，肥健不饥，轻身延年”译为 Long-term taking [of it will] tonify marrow, replenish Qi, strengthen the body, tolerate hunger, relax the body and prolong life。[①]

《神农本草经》的译者李照国指出，中医词语翻译主要有六大原则，即“自然性、简洁性、民族性、回译性、同一性、规定性。”[②]

“《本草纲目》是由明代医药学家李时珍编撰而成的一部医药百科全书。早在17世纪，在华的耶稣会士便开始将《本草纲目》作为研究中国博物学和医学的典籍文献，将其介绍到西方。17世纪前半叶，法国医生旺德蒙德（Jacques Francois Vander-monde，1723—1762）在澳门行医过程中，在当地人的帮助下，选译了《本草纲目》卷五至卷十一部分的内容，并编写了法文材料《〈本草纲目〉中水、火、土、金石诸部药物》(*Eaux, feu, terres, mineraux, metaux et sels du Pen-Tsau Kang-Mu*)。而第一个用西文公开出版的节译本，还要属18世纪的著名法国汉学家杜赫德在《中华帝国全志》中摘译的《本草纲目》部分内容，标题为“《本草纲目》摘录，中国植物志或中国医用博物学(*Extrait du Pentsao cang mou , c'es-a-dire, de l'herbier Chinois, ou Histoire naturelle de la Chine pour l'usage de laMedecine*)”。范德蒙德、杜赫德和伊博恩的相关著作对《本草纲目》的内容进行了部分翻译。至于英文全译本，至今为止仅有罗希文于2003年所译的1种。

---

① 孙星衍.神农本草经(汉英对照)[M].刘希茹，李照国，译.上海：上海三联书店，2017.

② 赵丽梅，汪剑.认知翻译学视角下《神农本草经》的英译研究[J].中国中医基础医学杂志，2022，28(8)：1335-1338.

第三节　植物典籍双语赏析

【赏析一】

《救荒本草》节选①

独扫苗

生田野中,今处处有之。叶似竹形而柔弱细小,抪茎而生。茎叶稍间结小青子,小如粟粒。科茎老时可为扫帚。叶味甘。

救饥:采嫩苗叶煠熟,水浸淘净,油盐调食。晒干煠食,不破腹尤佳。

治病:今人多将其子亦作地肤子代用。

Dusaomiao [独扫苗, kochia shoot, Kochia scoparia (L.) Schrad.]

Kochia scoparia, Dusaomiao,grows in the field and can be seen everywhere presently. Its leaves, scattered on the stems, are like the bamboo leaves, tender and smaller. Its fruit is cyan,small as a chestnut, growing at the end of branches. Its plants can be made into brooms when old and dry. Its leaves are sweet in taste.

For famine relief: Collect and blanch young plants and leaves, soak them in fresh water , wash and clean them, and flavor them with oil and salt. It can also betaken after being dried and blanched. It is better in quality if no diarrhea happens after taken.

For disease treatment: Its fruit is also regarded as Difuzi [地肤子, kochia, Kochiae Fructus] and is used as medicinal to replace it.

(范延妮 译)

① 朱橚.《救荒本草》汉英对照[M].范延妮,译.苏州:苏州大学出版社,2019.

紫云菜

生密县付家冲山野中。苗高一二尺。茎方，紫色，对节生叉。叶似山小菜叶，颇长，抪梗对生。叶顶及叶间开淡紫花。其叶味微苦。

救饥：采嫩苗叶煠熟，水浸淘去苦味，油盐调食。

Ziyuncai［紫云菜，peruvian groundcherry herb］Clinopodium chinense (Benth.)

Ziyuncai grows in the fields of Fujia chong in Mixian County. Its plants are 1–2 Chi high and its purple stems are square with opposite branches. Its opposite leaves are like those of platycodon but longer. Its flowers are light purple at the end of branches or axilla. Its leaves are slightly bitter in taste.

For famine relief: Collect and blanch young plants and leaves, remove the bitterness taste by washing them in fresh water, and then flavor them with oil and salt.

紫香蒿

生中牟县平野中。苗高一二尺。茎方紫色。叶似邪蒿叶而背白；又似野胡萝卜叶微短。茎叶稍间结小青子，比灰菜子又小。其叶味苦。

救饥：采叶煠熟，水浸去苦味，换水淘净，油盐调食。

Zixianghao［紫香蒿，Annuae Sweet Wormwood Herb, Artemisia］

Zixianghao grows in the flat fields of Zhongmu County. Its plants are 1-2 Chi high and its square stems are purple. Its leaves are like those of Xiechao [邪蒿, seseli, Seseli Herba], with the white backside, and also like those of Ye Huluobo [野胡萝卜, wild carrot, Daucus carota L.] but shorter. Its in florescences bear green fruit, smaller than that of wild pig weeds. Its leaves are bitter in taste.

For famine relief: Collect and blanch leaves, remove the bitterness by soaking them in fresh water, wash and clean them, and then flavor them with oil and salt.

(范延妮 译)

青荚儿菜

生辉县太行山山野中。苗高二尺许,对生茎叉,叶亦对生。其叶面青背白,锯齿三叉叶,脚叶花叉颇大,状似荏子叶而狭长尖艄。茎叶稍间开五瓣小黄花,众花攒开,形如穗状。其叶味微苦。

救饥:采嫩苗叶煠熟,换水浸,淘去苦味,油盐调食。

Qingjia'er Cai [青荚儿菜, heterophyllous Patrinia, Patrinia heterophylla Bunge]

Qingjia'er Cai grows in the mountain and wilderness of Taihang Mountain of Huixian County. Its plants are about two Chi high, with opposite branches and leaves. The upside of its leaves is green and backside white, saw-tooth-shaped and trifid. Its basal leaves are partite with the shape like that of Renzi [荏子, perilla, Perilla frutescens (L.) Britt.] but narrower, longer and pointed. Its yellow flowers are small and five-petaled on inflorescences, gathered and fringy. Its leaves are a little bitter in taste.

For famine relief: Collect and blanch young plants and leaves, soak them in fresh water, remove the bitterness by washing them, and then flavor them with oil and salt.

(范延妮 译)

回回米

本草名薏苡人,一名解蠡,一名屋菼,一名起实,一名赣,俗名草珠儿,又呼为西番蜀秫。生真定平泽及田野。交趾[1]生者子最大,彼土人呼为赣珠,今处处有之。苗高三四尺。叶似黍叶而稍大。开红白花，作穗子，结实青白色，形如珠而稍长，故名薏珠子。味甘，微寒，无毒。今人俗亦呼为菩提子。

救饥：采实，舂取其中人煮粥食。取叶煮饮亦香。

治病：文具《本草·草部》薏苡人条下。

Huihuimi [回回米, the seed of job's tears, Coix chinensis Tod.]

Huihuimi also named Yiyiren, Jieli, Wutan, Qishi, Gan, Caozhu'er and Xifan Shushu, grows in the flat swamps and fields of Zhending. The species with the biggest seeds which originate in Jiaozhi is called Ganzhu by local people, and presently it can beseen everywhere. Its plants are 3-4 Chi high. Its leaves are like but bigger than those of Shu [黍, millet, Panicum miliaceum L.]. The female spikelets are red while the male spikelets are yellow or white. The green-white grain is shaped like the bead but longer, so it is called Yizhuzi. It is sweet in taste, slightly cold in nature and non-toxic. Presently people always call it Putizi.

For famine relief: Collect grains, grind them and peel off the coat for cooking congee, or collect leaves, boil them for tea.

For disease treatment: See Yiyiren Clause in *Materia Medica · Herbaceous Plant*.

（范延妮 译）

Notes

[1] One variation of Yiyi imported from Vietnam.

【赏析二】

节选自《神农本草经》①

干漆

味辛温无毒。主绝伤补中,续筋骨,填髓脑,安五脏,五缓六急,风寒湿痹,生漆,去长虫。久服轻身耐老。生川谷。

【考据】

1.《名医》曰:生汉中,夏至后采,干之。

2. 案《说文》云:桼木汁可以鬃物,象形,鬃如水滴而下,以漆为漆水字;《周礼》载师云:漆林之征,郑元云:故书漆林为桼林;杜子春云:当为漆林。(孙星衍)

【今译】

味辛,性温,无毒

主治严重受伤,能补益人体内脏,能使跌打损伤中的筋骨续接,能填充脑髓,能安静五脏,能治疗小儿五迟和六种极度损伤以及风寒湿邪所致之痹证。生漆能祛除蛔虫。长期服用能使人身体轻盈,不易衰老,该物生长在山川河谷之中。

Gangi (干漆, dried lacquer of true lacquertree Resina Toxicodendri)[1]

Gangi, pungent in taste, warm and non-toxic [in property], [is mainly used] to treat severe damage, tonify the middle (internal organs), to remedy [severe injury of] sinews and bones, to enrich the brains, to harmonize the five Zang-organs, to treat five [kinds of] retardations[2], six [kinds of] extreme [syndromes/ patterns] [3] and wind-cold dampness impediment. Raw lacquer can kill

① 孙星衍.神农本草经(汉英对照)[M].刘希茹,李照国,译.上海:上海三联书店,2017.

roundworms. Long-term taking [of it will] relax the body and prevent aging. [It] grows in mountain valleys and river valleys.

[Textual Research]

In the book entitled] *Ming Yi Bie Lu* (《名医别录》, *Special Record of Great Doctors*), [it] says [that Ganqi (干漆,dried lacquer of true lacquer tree, Resina Toxicodendri)] grow: in Hanzhong (汉中) [It can be] collected in Summer Solstice and dried.

Notes

1. Ganqi (干漆, dried lacquer of true lacquertree, Resina Toxicodendri) is pungent in taste, warm and non-toxic in property,entering the liver meridian and the spleen meridian, effective in breaking blood stasis, resolving accumulation and killing worms. Clinically it is used to treat amenorrhea due to blood stasis, abdominal mass, abdominal pain due to blood stasis and retention of worms.

2. Five kinds of retardations include retardations in walking, speaking, standing, fontanel closure and tooth growth.

3. Six kinds of extreme syndromes/patterns include extreme disorders of Oi, blood, sinews, bones, muscles and essence.

(李照国 译)

禹余粮

味甘，寒。主咳逆寒热，烦满下(《御览》有痢字)，赤白，血闭，症瘕，大热。炼饵服之，不饥，轻身延年。生池泽及山岛中。

【考据】

1.《名医》曰：一名白余粮，生东海及池泽中。

2.案《范子计然》云：禹余粮出河东；《列仙传》云：赤斧，上华山取禹余粮；《博物志》云：世传昔禹治水，弃其所余食于江中，而为药也，按此出神农

经,则禹非夏禹之禹,或本名白余粮,《名医》等移其名耳。(孙星衍)

【今译】

味甘,性寒。主治咳嗽气逆,寒热性疾病,烦躁,郁闷,痢疾,赤白带下,经闭,症疲,身体高热。烧炼后服用,能使人减少饥饿之感,身体轻盈,延年益寿。该物生长在池泽及山岛之中。(刘希茹)

Yuyuliang (禹余粮, limonite, Limonitum)[1]

Yuyuliang, sweet in taste and cold [in property], [is mainly used] to treat cough with dyspnea, cold - heat [disease], vexation and fullness[2] red and white dysentery, amenorrhea, abdominal mass and severe heat[3] [disease]. Taking refined [Yuyuliang 禹余粮, limonite, Limonitum) will enable people] to tolerate hunger, relax the body and prolong life. [It] exists in lakes and swamps as well as mountains and islands.

[Textual Research]

In the book entitled *Ming Yi Bie Lu* (《名医别录》, *Special Record of Great Doctors*), [it] says [that Yuyuliang (禹余粮, limonite, Limonitum) is also] called Baiyuliang (白余粮), produced in Donghai (东海, East Sea) and Chize (池泽).

In the book entitled *Fan Zi Ji Ran* (《范子计然》, *Studies About Fan Li's Teacher*), [it] says [that] Yuyuliang (禹余粮, limonite, Limonitum) is produced in Hedong (河东). [In the book entitled] *Lie Xian Zhuan* (《列仙传》, *Story About Immortals*) , [it] says [that] Yuyuliang (禹余粮, limonite, Limonitum) can be collected from Huashan (华山) with red axe.

Notes

1. Yuyuliang (禹余粮, limonite, Limonitum) is a mineral medicinal, also called Yuliangshi (禹粮石) and Taiyi Yuyuliang (太乙禹余粮), sweet in taste

and astringent in property, entering the meridians of the spleen, stomach and large intestine, effective in ceasing diarrhea, hemorrhage and leucorrhea. Clinically it is used to treat chronic diarrhea due to deficiency-cold, chronic dysentery, bloody stool, uterine flooding and spotting as well as leucorrhea.

2. Fullness refers to discomfort and restlessness in the chest and heart.

3. Severe heat refers to excess-heat syndrome/pattern caused by exuberance of Yang.

（李照国 译）

菊花

味苦，平。主诸风头眩，肿痛，目欲脱，泪出，皮肤死肌，恶风湿痹。久服，利血气，轻身，耐老延年。一名节华，生川泽及田野。

【考据】

1.《吴普》曰：菊华一名白华（初学记），一名女华，一名女茎。

2.《名医》曰：一名日精，一名女节，一名女华，一名女茎，一名更生，一名周盈，一名傅延年，一名阴成，生雍州。正月采根，三月采叶，五月采茎，九月采花，十一月采实，皆阴干。

3. 案《说文》云：蘜治墙也，蘜日精也，似秋华，或省作，《尔雅》云，蘜治墙；郭璞云：今之秋华菊，则蘜、蘜、蘜皆秋华，字惟今作菊，《说文》以为大菊瞿麦，假音用之也。（孙星衍）

【今译】

主治各种风邪所致的头目眩晕，眼睛肿痛欲脱，泪流不止，皮肤因失去知觉而出现的类似死亡的肌肉，麻风病，湿邪所致痹证。长期服用，有利于血气的运行，能使人身体轻盈，不易衰老，能延年益寿。又称为节华，该物生长在山川和平泽之中及田野。（刘希茹）

Juhua (菊花, flower of florists chrysanthemum. Flos Chrysanthemi) [Original Text]

Juhua, bitter in taste and mild [in property], [is mainly used] to treat dizziness of head with swelling and pain, eyes tending to prolapse [due to distending pain], tearing, numbness of skin like withered muscle, aversion to wind and dampness impediment.

Long-term taking [of it will] promote circulation of blood and flow of Qi, relaxing the body, preventing aging and prolonging life. [It is also] called Jiehua (节华), growing in valleys, swamps and fields.

[Textual Research]

[In the book entitled] *Wu Pu Ben Cao* (《吴普本草》, Wu Pu's *Studies of Materia Medica*), [it] says [that] Juhua (菊花, flower of florists chrysanthemum, Flos Chrysanthemi) is also known as Baihua (白华), Nǚhua (女华) and Nǚjing (女茎).

[In the book entitled] *Ming Yi Bie Lu* (《名医别录》, *Special Record of Great Doctors*), [it] says [that Juhua (菊花, flower of florists chrysanthemum, Flos Chrysanthemi) is also] called Rijing (日精), Nǚjie (女节), Nǚhua (女华), Nǚhuajing (女华茎), Gengsheng (更生), Zhouying (周盈), Fuyannian (傅延年) and Yincheng (阴成), growing in Yongzhou (雍州). [Its] root [should be] collected in January, [its] leaves [should be] collected in March, [its] stalks should be] collected in May, [its] flowers [should be] collected in September and [its] fruits [should be] collected in November, all of which should be dried in the shade.

Notes

Juhua (菊花, flower of florists chrysanthemum, Flos Chrysanthemi) is a herbal medicinal, sweet and bitter in taste, cool in property, entering the liver meridian and lung meridian, effective in dispersing wind, pacifying the liver, improving vision and removing toxin. Clinically it is used to treat exogenous wind-heat disease, headache, dizziness, redness of the eyes, furuncle and swelling).

(李照国 译)

【赏析三】

《本草纲目》节选

大豆

尗俗作菽。〔时珍曰〕豆、尗皆荚谷之总称也。篆文尗，象荚生附茎下垂之形。豆象子在荚中之形。《广雅》云：大豆，菽也，小豆，荅也。角曰荚，叶曰藿，茎曰萁。

Dadou

Dadou, Black soybean, SHU (尗) or SHU (菽). Li Shizhen: Dou and Shu are generai names of beans. Shu written as a seal character symbolizes the pod hanging from the stem of the bean. Dou symbolizes the seed in the pod.

The book *Guang Ya*: Dadou is merely Shu and Xiaodou (Chixiaodou/semen phaseoil/rice bean) is Da. Its pod is called Jia. Its leaf is called Huo and its stem is called Qi.

曼陀罗花

风茄儿(《纲目》)、山茄子。

[时珍曰]法华经言佛说法时，天雨曼陀罗花。又道家北斗有陀罗星使者，手执此花。故后人因以名花。曼陀罗，梵言杂色也。茄乃因叶形尔。姚伯声花品呼为恶客。

【集解】

时珍曰：曼陀罗生北土，人家亦栽之。春生夏长，独茎直上，高四五尺，生不旁引，绿茎碧叶，叶如茄叶。八月开白花，凡六瓣，状如牵牛花而大。攒花中坼,骈叶外包,而朝开夜合。结实圆而有丁拐,中有小子。八月采花,九月采实。

【气味】

辛,温,有毒。

【主治】

诸风及寒湿脚气,煎汤洗之。又主惊痫及脱肛,并入麻药(时珍)。

【发明】

[时珍曰]相传此花笑采酿酒饮,令人笑;舞采酿酒饮,令人舞。予尝试之,饮须半酣,更令一人或笑或舞引之,乃验也。八月采此花,七月采火麻子花,阴干,等分为末。热酒调服三钱,少顷昏昏如醉。割疮炙火,宜先服此,则不觉苦也。

Mantuoluo Hua, Datura

Shanqiezi

Li Shizhen: According to *Fahua Jing*, a Buddhist sutra, when the Buddha was giving his lecture, it rained Mantuoluo hua. In Taosim, there is an immortal named Tuoluoxing Shizhe, who had this flower in his hand. The name of the drug is given in accordance with the immortal. Mantuoluo is a transliteration of a Sanskrit term, "multi-colored." As to the origin of such names as Feng qie'er and Shanqiezi (both are related to eggplant), it is because the leaf of this herb is similar to Qieye/folium solani melongenae/eggplant leaf. In Yao Bosheng's book *Hua Pin* (*Collections of Flowers*), the drug is called Eke.

[Previous Explanations]

Li Shizhen: Mantuoluo grows in the northern part of the country. It is also cultivated by farmers. It begins to grow in spring and develops in summer. The only stem grows upward four to five chi tall. There are no forks. The stem and leaf are green. The leaf is similar to Qieye. A white flower blooms in the eighth month with six petals. It looks like Qianniuhua/flos pharbitidis/pharbitis flower, but is bigger. Flocks of stamens and pistils gather in the center of the flower, which is covered with surrounding leaves. The flower blooms in the day and closes at night. The fruit is round, with a handle, and there are small seeds inside. Collect the flower in the eighth month and the fruit in the ninth month.

[Quality and Taste]

It is pungent, warm and toxic.

[Indications]

Li Shizhen: It is good for dispersing the invading pathogenic Wind and treat beriberi due to invasion of pathogenic Cold and Humidity. Stew the drug in water to make a decoction and wash the affected part. It is also good for treating convulsion and epilepsy and a prolapsed rectum. It is used in anesthetics.

[Explication]

Li Shizhen: An anecdote says that, when the drug is collected by a person who is laughing and then is brewed with wine, drinking of it will induce the person to laugh. And, when the drug is collected by a person who is dancing and then is brewed with wine, drinking of it will make the person dance. I tried. After drinking such wine to one ' s content, and if there is another person who is laughing or dancing in front of the drinker, the drinker will laugh and dance, too. Collect the flower in the eighth month. Collect Huomazihua/flos cannabis/hemp flower in the seventh month. Dry the drugs in shade. Grind equal amounts of the two flowers into powder and take three qian with hot wine. Shortly afterwards, the person will get drunk and tend to sleep. When a sore is to be dissected and moxibustion is to be practiced, administer this to the patient. In this way, he will not feel the pain.

(罗希文 译)[①]

荞麦

【释名】

荍(音翘)麦、乌麦(吴瑞)。花荞。

[时珍曰]荞麦之茎弱而翘然,易长易收,磨面如麦,故曰荞曰荍,而与麦同名也。俗亦呼为甜荞,以别苦荞。杨慎丹铅录,指乌麦为燕麦,盖未读日用本草也。

【集解】

[炳曰]荞麦作饭,须蒸使气馏,烈日曝令开口,舂取米仁作之。

[时珍曰]荞麦南北皆有。立秋前后下种,八、九月收刈,性最畏霜。苗高一、二尺,赤茎绿叶,如乌桕树叶。开小白花,繁密粲粲然。结实累累如羊蹄,实有三棱,老则乌黑色。王祯农书云:北方多种。磨而为面,作煎饼,配蒜食。或作汤饼,谓之河漏,以供常食,滑细如粉,亚于麦面。南方亦种,但作粉饵食,乃农家居冬谷也。

【气味】

甘,平,寒,无毒。

[思邈曰]酸,微寒。食之难消。久食动风,令人头眩。作面和猪、羊肉热食,不过八、九顿,即患热风,须眉脱落,还生亦希。泾、邠以北,多此疾。又不可合黄鱼食。

---

① 李时珍. Condensed Compendium of Materia Medica. Vol. Ⅰ, Ⅱ, Ⅲ, Ⅳ, Ⅴ. 本草纲目选(全五册)[M]. 罗希文,译. 北京:外文出版社,2012.

【主治】

实肠胃，益气力，续精神，能炼五脏滓秽（孟诜）。作饭食，压丹石毒，甚良萧炳。以醋调粉，涂小儿丹毒赤肿热疮（吴瑞）。降气宽肠，磨积滞，消热肿风痛，除白浊白带，脾积泄泻。以沙糖水调炒面二钱服，治痢疾。炒焦，热水冲服治绞肠沙痛时珍。

【发明】

[颖曰]本草言荞麦能炼五脏滓秽。俗言一年沉积在肠胃者，食之亦消去也。

[时珍曰]荞麦最降气宽肠，故能炼肠胃滓滞，而治浊带泄痢腹痛上气之疾，气盛有湿热者宜之。若脾胃虚寒人食之，则大脱元气而落须眉，非所宜矣。孟诜云：益气力者，殆未然也。按杨起简便方云：肚腹微微作痛，出即泻，泻亦不多，日夜数行者。用荞麦面一味作饭，连食三四次即愈。予壮年患此两月，瘦怯尤甚。用消食化气药俱不效，一僧授此而愈，转用皆效，此可征其炼积滞之功矣。普济治小儿天吊及历节风方中亦用之。

Qiaomai

[Explanation of Names]

Qiaomai (荞麦)

Wumai-Wu Rui.

Huaqiao

Li Shizhen: Qiaomai has weak, but upright stem. It is a herb that grows easily and can be harvested conveniently. The seed can be ground into flour, which is similar to Xiaomai/semen tritici/wheat. It is colloquially called Tianqiao (meaning "sweet Qiao") to differentiate from Kuqiaomai (meaning "bitter Qiao"). Yang Shen in his work *Danqian Lu* said that Wumai was just Yanmai (Quemai/

herba bromi japonica/herb of Japanese bromegrass), but he was wrong, he did not read of the book *Riyong Bencao* (*Materia Medica for Daily Use*).

[Previous Explanations]

Xiao Bing: First steam buckwheat and then dry it in the sun until it cracks. Husk the thing to get the grain, and then it can be used to make food.

Li Shizhen: Qiaomai is found both in the south and north. Sow the seed around the day of the solar term of the Beginning of Autumn (August 7). Harvest the crop in the eighth and ninth months. It is a herb that is easily damaged by frost. The seedling is one or two *chi* tall. The stalk is red and the leaf is green, and it looks like Wujiuye/folium sapii radicis/leaf of Chinese tallow tree. Small white flowers bloom and gather together. There is abundant fruit similar to Yangti/frutus rumicis japonica/fruit of Japanese dock. The seed is triangular in shape, and it turns black when ripe.

The book *Nong Shu* by Wang Zhen: The herb is planted in bulk in the north. People grind the grain into flour and make stir-fried cakes eaten together with garlic. Or, it can be made into gruel. It is locally called Helou. It is a common food. It is fine and slippery like rice powder. It is secondary to Maimian/semen tritici/wheat flour. The herb is also planted in the south. People prepare noodles with it. It is a common food in winter for farmers.

[Quality and Taste]

It is sweet, plain, cold and nontoxic.

Sun Simiao: It is sour and slightly cold, and is hard to digest. Long-term use of it will stir up pathogenic Wind and cause vertigo. People eat hot noodles made of Qiaomai with pork and mutton. But, after consuming eight or nine portions of such noodles, the person will suffer from febrile disease caused by pathogenic

Heat with his beard and eyebrows falling off.It is very difficult case to overcome. People in Jing and Bin areas suffer such a disease quite often. Also, this food should not be served together with Huangyu/huso dauricus/huso sturgeon.

[Indications]

Meng Xian: It fills the intestine and stomach, reinforces qi and bodily strength and enhances one's spirit. It refines the sediment and residue of the Five Viscera.

Xiao Bing: Steam the drug and eat it to suppress toxin of alchemical stone drugs. It is very effective.

Wu Rui: Blend the flour of the drug with vinegar. Apply it to treat infantile erysipelas with red and swollen skin and sore with hot feeling.

Li Shizhen: It brings down adverse ascending gas and facilitates bowel movement. It helps dissolve indigestion and stagnation. It eliminates hot swelling and arthralgia due to invading pathogenic Wind. It eliminates whitish and turbid urine and treats leucorrhea. It stops diarrhea due to Spleen accumulation. Blend two *qian* of stir-fried flour of Qiaomai with sugared water. Take it to stop dysentery. Stir-fry the flour of Qiaomai until it becomes scorched. Wash it down with hot water to relieve the pain in dry cholera (or exanthema of intestine).

[Explication]

Wang Ying: *Shen Nong Bencao Jing* (*Shen Nong's Great Herbal*) says that the drug is good to refine the sediment and residue resting with the Five Viscera. This can be understood as meaning that the indigestion accumulated in the Stomach and Intestine in the last year can be cleared away.

Li Shizhen: Qiaomai is a drug that brings down the adverse ascending of gas and facilitates bowel movement. So, it is good to refine the sediment and residue resting in the Stomach and Intestine. It is, therefore, effective to treat diarrhea and dysentery with stagnation of turbidity and adverse ascending of gas. It is especially good for people whose qi is rampant and with invading pathogenic Humidity and Heat. If a person suffering from a debilitated and cold condition of the Spleen and Stomach takes the drug, his Yuanqi (Primordial Vital Energy) will be badly consumed and his beard and eyebrows will fall off. It is strictly prohibited. Meng Xian said that the drug "reinforces qi and strength," but this is not correct.

The book *Jianbian Fang* by Yang Qi: When the patient suffers from a vague pain in the abdominal region and produces a stool, although only in a small amount, several times in the day or night when he moves around, the treatment is to cook food with flour of Qiaomai. After eating three or four meals made of Qiaomai, the syndrome will be gone. When I was in my middle age, I suffered from this disease for two months. Drugs designed to dissolve indigestion and to facilitate the function of qi were prescribed, but they did not work. One day, a monk told me of the above treatment. I treated myself accordingly and recovered. From then on, I used the prescription on many other people and it worked every time. From this, we can see its function of dissolving indigestion and stagnation. In *Puji Fang* (*Prescriptions for Universal Relief*), the drug is also used in prescriptions designed to treat infantile convulsion with uplifted eye and arthritis.

(罗希文 译)[①]

---

① 李时珍. Condensed Compendium of Materia Medica. Vol. Ⅰ, Ⅱ, Ⅲ, Ⅳ, Ⅴ. 本草纲目选(全五册)[M]. 罗希文,译. 北京:外文出版社,2012.

莲藕(本经上品)

【释名】

其根藕(《尔雅》)。其实莲(同上)。其茎叶荷。

[时珍曰]尔雅以荷为根名,韩氏以荷为叶名,陆玑以荷为茎名。按茎乃负叶者也,有负荷之义,当从陆说。乃嫩蒻,如竹之行鞭者。节生二茎,一为叶,一为花,尽处乃生藕,为花、叶、根、实之本。显仁藏用,功成不居,可谓退藏于密矣,故谓之蔤。花叶常偶生,不偶不生,故根曰藕。或云藕善耕泥,故字从耦,耦者耕也。茄音加,加于蔤上也。蕸音遐,远于密也。菡萏,函合未发之意。芙蓉,敷布容艳之意。莲者连也,花实相连而出也。菂者的也,子在房中点点如的也。的乃凡物点注之名。薏犹意也,含苦在内也。古诗云:食子心无弃,苦心生意存。是矣。

【集解】

[时珍曰]莲藕,荆、扬、豫、益诸处湖泽陂池皆有之。以莲子种者生迟,藕芽种者最易发。其芽穿泥成白蒻,即蔤也。长者至丈余,五六月嫩时,没水取之,可作蔬茹,俗呼藕丝菜。节生二茎:一为藕荷,其叶贴水,其下旁行生藕也;一为芰荷,其叶出水,其旁茎生花也。其叶清明后生。六七月开花,花有红、白、粉红三色。花心有黄须,蕊长寸余,须内即莲也。花褪连房成菂,菂在房如蜂子在窠之状。六七月采嫩者,生食脆美。至秋房枯子黑,其坚如石,谓之石莲子。八九月收之,斫去黑壳,货之四方,谓之莲肉。冬月至春掘藕食之,藕白有孔有丝,大者如肱臂,长六七尺,凡五六节。大抵野生及红花者,莲多藕劣;种植及白花者,莲少藕佳也。其花白者香,红者艳,千叶者不结实。别有合欢(并头者),有夜舒荷(夜布昼卷)、睡莲(花夜入水)、金莲(花黄)、碧莲(花碧)、绣莲(花如绣)。皆是异种,故不述。相感志云:荷梗塞穴鼠自去,煎汤洗镴垢自新。物性然也。

Lian'ou

[Explanation of Names]

Ou (root)

–*Er Ya*

Lian (fruit)

–*Er Ya*

He (rhizome and leaf)

Li Shizhen: In *Er Ya,* He is the name of its root (rhizome). But Han Baosheng said that He is the name of the leaf. Lu Ji called the stalk He. Li Shizhen's comment: As the stalk carries the leaf, so it is correct to call the stalk He. This is because the character He means "carry". So Lu Ji was correct. Mi is the tender stalk, In each stalk, there is one leaf and the other is the flower. To the far end at the bottom, the rhizome develops, which is the base of its flower, leaf, root and fruit. The plant hides itself, so it is called Mi (to hide oneself in secrecy). The flower and leaf are always growing in pairs, so it is called Ou ("double"). Some say the name Ou is due to the fact that the rhizome digs deeply into the mud, so it is called Ou (also meaning "plowing"). Jia (stalk) is above the Mi. Xia (leaf) is far away from the Mi. Handan (flower bud) means something closed but ready to grow, Furong are characters describing the beautiful petals stretching about. Lian means the fruit is connected with the flower. Di is the seed. Yi is the bitter bud inside the seed. An ancient poem goes, "Do not abandon the heart of the lotus seed. Bitter as it is, it embraces a living tendency."

[Previous Explanations]

Li Shizhen: Lotus is found in ponds and marshes in Jingzhou, Yang zhou, Yuzhou and Yizhou. If it is planted by sowing the seed, it grows slowly. It is easier for the plant to develop if the bud of the lotus rhizome is planted. Its bud develops into a white stalk and grows through the mud out of the water. This is Mi, which can grow ten *chi* or more. In the fifth and sixth months, when the stalk is still tender, people dive into water and collect it. The thing is called Ousicai, which can be served as a vegetable. Two stems grow on each node of the stalk. One stem has a leaf above the water and the lotus rhizome is developed in the mud. The other stem has a leaf and a flower. The leaf of lotus begins to grow after the day of the solar term of the Pure Brightness (April 5). The flower blooms in the sixth and seventh months. There are three flower colors: red, white and pink. In the center of the flower there are yellow tassels. The pistil is half a *cun* long.

Within the tassels is the flower. After the flower withers, seeds are formed in the seedpod. The seeds and its seed pots look like the nest and its bees. In the sixth and seventh months, collect the tender seedpod. The tender seed is crispy and delicious. In autumn, the seedpod withers and the seed becomes black and hard as stone. So, it is called Shilianzi ("stony lotus seed"). Collect the seed in the eighth and ninth months. After peeling off the black capsule, the seed is sold in all places. It is called Lianrou ("lotus seed pulp"). People dig out the lotus rhizome in winter or spring and serve it as a vegetable. The rhizome is white with holes and fibers inside. The big one can be as big as an arm. It can be six to seven *chi* long with five to six segments. Generally speaking, the wild one and that with red flower has an inferior lotus rhizome, but with more lotus seed. The domestic one, or that with a white flower, has less seed, but a superior rhizome. The white flower is fragrant. The red flower is beautiful; that with multi petals does not bear seed. There are several varieties of lotus:

—HEHUAN (with double flowers);

—YESHUHE: the leaf spreads about at night but curls up in the day,

—SHUILlAN: the flower enters the water at night;

—JINLIAN: with a yellow flower;

—BILIAN: with a green flower,

—XIULIAN: flower with designs as if embroidered.

All these are varieties and although they are not explained here.

The book *Xianggan Zhi*: "Insert a lotus stalk in a cave, and all rats will flee away. Scalded pots can be washed anew with a decoction of lotus leaf."

(罗希文 译)①.

练　习

1.句子翻译。

1)白芥子,粗大白色,如白粱米,甚辛美,从戎中来。(《本草纲目》)

2)白芥生太原、河东。叶如芥而白,为茹食之。(《本草纲目》)

3)茎叶间开五瓣大黄花。结瓜,形如黄瓜而大,色青,嫩时可食,老则去皮,内有丝缕,可以擦洗油腻器皿,味微甜。(《救荒本草》)

4)*The Complete Treatise on Agriculture* is a Chinese agricultural text compiled by Xu Guangqi in the early 17th century. It covers diverse topics such as land reclamation, irrigation, tools, sericulture, and famine relief, drawing upon over 300 earlier works. (*The Complete Treatise on Agriculture*)

2.段落练习。

1)《茶谱》:"衡州之衡山、封州之西乡,茶研膏为之,皆片团如月。又彭州蒲

① 李时珍. Condensed Compendium of Materia Medica. Vol. Ⅰ, Ⅱ, Ⅲ, Ⅳ, Ⅴ. 本草纲目选(全五册)[M]. 罗希文,译.北京:外文出版社,2012.

村、堋口,其园有'仙芽'、'石花'等号。"明人《月团茶歌序》:"唐人制茶碾末,以酥滫为团,宋世尤精。元时其法遂绝。予效而为之,盖得其似,始悟唐人咏茶诗所谓'膏油首面',所谓'佳茗似佳人',所谓'绿云轻绾湘娥鬟'之句。饮啜之余,因作诗记之,并传好事。"(《续茶经》(茶之源))

2)据《月令广义》记载:"炒茶时一锅不能超过半斤,先干炒,而后微微洒上点水,再用布卷起来揉。""挑摘洗净的茶叶,先稍微蒸一下,等颜色变了再摊开,用扇子扇去湿热气。揉好之后,再用火烘焙干,用箬叶包起来。有人说:'会蒸青不如会炒青,会晾晒不如会烘焙。'这说明茶叶炒后烘焙乃是最佳方法。"

3)书中关于茶(他们叫干草叶子)的记载,可见阿拉伯国家当时还没有喝茶的习惯。书中记述:"中国国王本人的收入主要靠盐税和泡开水喝的一种干草税。在各个城市里,这种干草叶售价都很高,中国人称这种草叶叫'茶',这种干草叶比苜蓿的叶子还多,也略比它香,稍有苦味,用开水冲喝,治百病。"(《山海经》前言)

4) In ancient China, *Illustrated Catalogue of Plants* (《植物名实图考长编》) is a botanical monograph or flora of medicinal plants with high scientific values, and its compiling system is quite different from those of Chinese materia medica masterpieces in the past generations and actually belongs to the category of botany. The book was proofread and published in 1848, one year after the death of Wu Qijun (吴其濬), the author of the book.The book has 71,000 words and 38 volumes in total, covering 1,714 different kinds of plants with detailed description of their names, shapes, colors, scents, varieties, life habits and uses, as well as more than 1,800 pictures attached. The book is also featured by its abundant texts complete with illustrations based on field observation and supplementary with document references. The author had

consulted farmers and peasants in an open mind wherever he came across a problem about the plants in field investigation and drew pictures of the plant samples he had collected, which could also be used as important evidence for the identification of the family and genus of plants today. The book is compiled mainly based on material object observation, rather accurate and delicate at that time as a kind of plant illustration and marked a big progress in drawing of material objects. The book not only provides detailed description of the properties, scents, curative effect and usages of plants recorded in it from the perspective of pharmacology, but also examines and corrects the categories of different plants, especially those with the same names but refer to different plants and those with different names but refer to the same plant. Additionally, the book also mentions their morphologies, life habits, uses and places of origin. The records of plants in the book are not limited to the description of their medical properties and uses but also involve flora of medicinal plants. It is the first large flora of medicinal plants in China.

In 1884, Japan copied the book for the first time. Keisuke Ito spoke highly of the book in the preface he wrote for the book. In 1940, Tomitaro Makino from Japan drew a great deal from the book to compile his Illustrated Handbook of Japanese Plants. Besides, Loaf Mirri and Walker from the United States also have ever quoted and held in esteem the book in their masterpieces. *Illustrated Catalogue of Plants* is the first large flora of medicinal plants in China, and it has been kept in many countries all around the world in their national libraries today.

5)　农业高科技主要包括生物工程、信息技术、新材料和设备工程方面的新技术。农业生物工程技术主要包括基因工程、农作物分子标记、动物克隆和转基因技术。中国的农业基因工程研究起步于20世纪80年代初，80年代中期被列为“国家高科技研究和发展项目（即863项目）”。到1996年底，正在研究中的转基因植物总数已达到47种，基因总数达到103种。经过十年的努力，在基因工程的安全管理、抵抗病虫害、对除草剂免疫和改良农作物品种等方面取得的成绩令人鼓舞。

从20世纪80年代中期开始，国家在农业生物技术的关键领域增加了投入。通过利用丰富的生物基因库，并将传统工艺同高新技术紧密结合，加强了农业基础研究。总体科研水平有了提高。

比如，在水稻和大麦的基因分裂、专用基因的克隆和染色体绘图方面取得了长足的进步。华中农业大学对水稻杂交的优势作了基本的生物研究，并取得了新的突破。在预防和控制植物病虫害的基础研究方面，中国不仅对引起大麦黄矮病的CPV病毒进行了基因排序分析，而且克隆了表层蛋白质基因。

20世纪90年代初，中国科学家开始对农作物基因组及其相关领域进行研究，这比国外同行晚了三到五年。1992年，一个农作物基因组研究项目——中国水稻基因组项目——正式启动。此后，许多和农作物基因组有关的研究项目纷纷被列为国家级、部级和地方级的科研项目。

拓展知识

1. 石声汉

石声汉（1907—1971），湖南湘潭人，1924年入武昌高等师范生物系，1933年赴英国伦敦大学求学，获植物生理学哲学博士学位，回国后历任西北农学院（今西北农林科技大学）、同济大学、武汉大学教授。

石声汉是系统今释并译介中国农学典籍的第一人，为中国古农学研究和传统农业科技文化“走出去”做出了杰出贡献。他的农学典籍翻译包含语内翻译和语际翻译两个维度。语内翻译以“校–注–释”为体例,充分尊重“原作–读者”:语际翻译属自译,运用薄译、译述结合、强制显化、语用显化等策略并调整编排和叙事逻辑,求真似、译精义、不多计工拙。其农学典籍译介模式集“释–著–译”三位一体,以“研而著”为基石,以服务读者为主要指导思想,对科技典籍翻译的底本选择、翻译策略和译才培养具有一定启示价值。

石声汉校注、今释农典以及出版农学研究专著共计十余部,发表农学研究论文数十篇,在中国开创了研究古农学的现代科学范式;自译农典研究论著《从齐民要术看中国古代的农业科学知识》(下文简称《从》)和《氾胜之书今释》,成为系统今释并译介中国农典的第一人。他为中国古农学研究和传统农业科技文化“走出去”做出了杰出贡献。汤佩松盛赞其农典研译工作广泛全面、深入系统,具有中国特色、达到国际水平,贡献卓越。据林广云等统计,其英译的《从》和《氾胜之书今释》分别传播到了12个和14个海外国家,是目前唯一的英文版本。石声汉的农典研译成果也得到了国外汉学家的高度评价,如:英国科技史专家李约瑟称其英译的《从》和《氾胜之书今释》已让他驰名西方学界;日本农史专家渡部武指出与其《齐民要术今释》“相融前后,日本的西山武一、熊代幸雄两先生的《校订注齐民要术》(上下两册,东京大学出版会,1957年、1959年)也出版发行”,并感佩其科学精神和农史贡献。概言之,石声汉的农典研译不仅促进了国内古农学的研究和农史学科的发展,也促进了国外汉学家对中国农业文化的研究和对中国农学典籍的翻译,提升了中国学术的国际地位。

1962年5月5日,西北农学院石声汉教授完成《齐民要术今释》,全书共97万多字。

石声汉长期从事生物学和植物生理学的教学与研究,是最早用科学方法研究中国哺乳类动物的学者之一,1951年后任西北农学院教授、古农学研究室主任，致力于整理、研究中国古代农业科学遗产工作，先后完成《齐民要术今释》《农政全书校注》等14部巨著，是我国著名的生物学家、植物生理学家和古农学家，同时也是中国农史学科重要奠基人之一。

《齐民要术》是由后魏贾思勰所著，是我国现存一部最古最完整的农书，有"中国古代农业百科全书"之谓，其也是全世界最古老的农业专著之一。该书印证经、史、子、集等书近200种，内容精湛丰富，但一千年来由于转抄传刻，混进了许多错、讹之字，成了一部难读的书。

石声汉年轻时翻看《齐民要术》，被一些古奥的文词和奇字所阻，未敢通读。几年后硬读一遍，更觉这部书的可贵，当时便希望能有对古农学有素养的有志之士，把这部奇书好好整理一番，让大家都能读懂。当时几个国家的学者都在动手研究，而且讥笑中国人不研究"贾学"是一件憾事，石声汉愤慨那些鄙视农圃、看不起自己祖国、自甘于"数典忘祖"而高谈阔论的"鸿儒"，于是依然投入到"古农学"这门"冷门"的学科上。

石声汉的工作效率非常惊人，只用三年工夫，就写了《齐民要术今释》97万字，《氾胜之书今释》5.8万字，《从齐民要术看我国古代农业科学知识》7.3万字，同时把后两书翻译成英文本，由科学出版社出版，在国外发行，再版四次，影响极大。此外他还写了8篇论文。

石声汉研究《齐民要术》的一些文章陆续发表后，引起了许多外国学者的重视。日本著名汉学家和中国农业科学史专家西山武一、天野元之助等六七位专家，都先后主动和石声汉建立了联系。日本研究《齐民要术》的权威西山武一教授看了《齐民要术今释》一、二分册后，赞叹为"贾学之幸"。他写信给石声汉，告知他已在和熊代幸雄共同翻译的日译本《齐民要术》上册的结尾处郑重声明"取消从前所说中国没有人研究《齐民要术》的话"。

西山武一还提出成立"中日研究《齐民要术》委员会"，地址设在西北农学院，后来因为种种原因未能实现来中国的愿望，来信表示莫大遗憾。熊代幸雄在来信中写道:“当我拿到盼望的贵著之后,高兴得几乎要跳起来。昨天,我花了一天的时间,初略地拜读了贵著,激动地我忘记了时间的流逝,给我的印象使我终生难忘。”

英国皇家科学院院士、著名的中国科学史专家李约瑟博士很早就和石声汉相识,信件往来更为密切,对《齐民要术今释》也极其肯定和重视。他认为:“由于他(石声汉)的两本著作——一本是关于前汉的农书作者氾胜之,另一本是关于六朝时期北魏贾思勰的不朽名著《齐民要术》的,他在西方世界已经很出名。”李约瑟在《中国科学技术史》“农业史”“生物史”两卷的扉页上写着“献给陕西武功张家岗西北农学院的石声汉教授”,并在后来的一封信中说“中国科学史农业卷的工作,极大地得益于石(声汉)先生的帮助”。

石声汉研究古农学的过程曲折而艰辛。1958年,《齐民要术今释》完稿时,石声汉遇上“反厚古薄今、反复辟倒退”的极左思潮,受到了错误批判,古农学研究室停开,研究工作受挫。然而批判过后,石声汉并没有因此裹足不前,而是又立刻投入《农政全书》的研究工作。世事无常,《农政全书校注》刚刚完成,石声汉又遇上了十年浩劫,被无故打成“牛鬼蛇神”。但他白天被轮流批斗,晚上仍笔耕不辍。工作成了他的习惯,是他的第一生命。

石声汉在古农学研究过程中还需与疾病作斗争。哮喘病、肺气肿和心脏病时时折磨着他,尤其到了冬季,他只能伏在桌上或床上拼命喘气。但是只要呼吸稍微舒畅些,他马上又伏案工作,经常一写就是几个小时,每天晚上都要熬到午夜两三点,一个月中还要熬几个通宵。

十年浩劫时,石声汉仅59岁,正是学术上出成果的黄金时代,但运动剥夺了他工作的权利,健康受到严重摧残。1971年春,石声汉腹痛发作,诊断为晚期胰腺癌。众所周知,胰腺癌是极其痛苦的,但石声汉以顽强的毅力忍受着巨大的病痛,即使痛得豆大的汗珠不停地冒,也从不哼一声,甚至在弥留之际,他仍惦记着工作。石声汉日以继夜地拼命工作,除了授课和培养研究生外，还完成了200余万字的《农政全书校注》《农桑辑要校注》《中国农业遗产要略》和《中国古代农书评介》等。在整个校勘过程中，石声汉呕心沥血，表现了高度负责的精神。有时为了一个疑难句或字，他往往要花费四五天时间，甚至查阅上百本书。

2. 节选《续茶经》(茶之造)——陆廷灿

据《北苑别录》："御园四十六所，广袤三十余里。自官平而上为内园，官坑而下为外园。方春灵芽萌坼，先民焙十余日，如九窠、十二陇、龙游窠、小苦竹、张坑、西际，又为禁园之先也。而石门、乳吉、香口三外焙，常后北苑五七日兴工。每日采茶、蒸，以其黄悉送北苑并造。"

"造茶旧分四局。匠者起好胜之心，彼此相夸，不能无弊，遂并而为二焉。故茶堂有东局、西局之名，茶有东作、西作之号。凡茶之初出研盆，荡之欲其匀，揉之欲其腻，然后人圈制銙，随笪过黄有方。故銙有花銙，有大龙，有小龙，品色不同，其名亦异，随纲系之于贡茶云。"

"采茶之法，须是侵晨，不可见日。晨则夜露未晞，茶芽肥润。见日则为阳气所薄，使芽之膏腴内耗，至受水而不鲜明。故每日常以五更挝鼓集群夫于凤凰山，[山有伐鼓亭，日役采夫二百二十二人。]监采官人给一牌，入山至辰刻，则复鸣锣以聚之，恐其逾时贪多务得也。大抵采茶亦须习熟，募夫之际必择土著及谙晓之人，非特识茶发早晚所在，而于采摘亦知其指要耳。"

"茶有小芽，有中芽，有紫芽，有白合，有乌蒂，不可不辨。小芽者，其小如鹰爪。初造龙团胜雪、白茶，以其芽先次蒸熟。置之水盆中，剔取其精英，仅如针小，谓之水芽，是小芽中之最精者也。中芽，古谓之一枪二旗是也。紫芽，叶之紫者也。白合，乃小芽有两叶抱而生者是也。乌蒂，茶之蒂头是也。凡茶，以水芽为上，小芽次之，中芽又次之。紫芽、白合、乌蒂，在所不取。使其择焉而精，则茶之色味无不佳。万一杂之以所不取，则首面不均，色浊而味重也。"

“惊蛰节万物始萌。每岁常以前三日开焙,遇闰则后之,以其气候少迟故也。”

“蒸芽再四洗涤,取令洁净,然后入甑,俟汤沸蒸之。然蒸有过熟之患,有不熟之患。过熟则色黄而味淡,不熟则色青而易沉,而有草木之气。故惟以得中为当。”

“茶既蒸熟,谓之茶黄,须淋洗数过,[欲其冷也。]方入小榨,以去其水,又入大榨,以出其膏,[水芽则以高榨压之,以其芽嫩故也。]先包以布帛,束以竹皮,然后入大榨压之,至中夜取出揉匀,复如前入榨,谓之翻榨。彻晓奋击,必至于干净而后已。盖建茶之味远而力厚,非江茶之比。江茶畏沉其膏,建茶惟恐其膏之不尽。膏不尽则色味重浊矣。”

“茶之过黄,初入烈火焙之,次过沸汤爁之,凡如是者三,而后宿一火,至翌日,遂过烟焙之,火不欲烈,烈则面泡而色黑。又不欲烟,烟则香尽而味焦。但取其温温而已。凡火之数多寡,皆视其銙之厚薄。銙之厚者,有十火至于十五火。銙之薄者,六火至于八火。火数既足,然后过汤上出色。出色之后,置之密室,急以扇扇之,则色泽自然光莹矣。”

【今译】

《北苑别录》记载:“皇家御用茶园共有四十六处,方圆三十多里地,从官平往上为内园,官坑往下的是外园。御园的茶春季比民间烘焙要早十几天,而九窠、十二陇、龙游窠、小苦竹、张坑、西际这些茶坊,在皇家御用茶园中又排在前面。而石门、乳吉、香口三处民间外焙,常常在北苑之后五、七天开工。每天采茶、蒸榨,处理好以后送到北苑去一块儿制造。”

“以前采制茶叶的茶坊分为四大部分。茶匠们在好胜心理驱使下相互攀比、彼此不服,因此常产生一些问题,于是就又把这个部分合成了两大处。这也就是茶堂里有东局和西局说法的来由,茶饼也因此分称东作、西作。刚出研盆的茶末都须搅拌均匀,揉搓细滑,然后再放进跨圈中模制,放在席架上晾晒至呈黄色为止。圈有花銙、大龙、小龙等，品第不同，名称也各异，都是按照贡茶的要求制作的。”

“采茶的时间须在凌晨，太阳出来即止。由于夜间凝结的露珠一大早尚未蒸发掉，茶芽肥硕而润泽。一旦受到阳光的灼射，茶芽里的内蕴就会消耗，制成的茶烹泡后颜色难保鲜亮。所以采茶季节一到，每天五更时分，一群茶人便到凤凰山（山上有伐鼓亭，每天需要二百二十二人）集合，监采官发给每人一个牌子入山采茶，到了日出须停采的时辰，监采官就鸣锣召回采茶人，唯恐大家一味贪多，导致超时滥采。大多数的采茶者都熟谙采茶之道，招募时就特别精心选择当地懂茶的人士，他们对有关茶的天时地情都十分了解。”

“茶的叶子分小芽、中芽、紫芽、白合、乌蒂，必须认真分辨。小芽形如小巧的鹰爪。当初造龙团胜雪和白茶最初用的就是小芽。先将茶芽蒸熟，再放进水盆中，挑选出其中的精华，针一般细小，被称为水芽，是小芽中最好的品第。中芽，古代时称作一枪二旗。紫芽叶子呈紫色。而白合，则是两片抱着小芽生长的托叶，乌蒂是茶叶的蒂。在所有形态的茶叶中，水芽是最上乘的，小芽稍微差一点，中芽又逊一筹。紫芽、白合、乌蒂，就根本要不得了。只要精心挑选，茶的色、味就绝对差不了。但如果挑拣不净，茶饼的表面就会不均匀，茶色浑浊，味道冲呛。”

“惊蛰时节，万物复苏。通常每年在惊蛰前两、三天开始焙茶如果逢遇闰年，节气推迟，采制工作也要相应往后略延。”

“准备蒸制的茶芽要反复洗涤，直至彻底清洁干净，再放进称作甑的蒸锅里，水沸后开蒸。蒸芽这道工序，既怕蒸过火，亦怕蒸不透。过熟则颜色变黄味道寡淡，不熟则颜色泛青容易沉淀，而且带有草木的生青气。所以火候掌握的合适很关键。”

“蒸熟的茶叶称为茶黄，茶黄必须先要淋洗几遍，（使之变冷）后放进小榨里挤压,目的主要是去掉中间的水分,再放进大榨里面把汁液榨出来。(水芽就用高榨压,因为它的芽比较嫩。)榨的时候先用布帛包起来,用竹皮捆绑,然后放进大榨里面压着,到半夜再取出来揉均匀,重新放进榨里面再榨,叫作翻榨,通宵击打,直到完全榨干。建茶味道厚重,不是江南一带的茶可相比的。江南茶怕榨干,建茶则怕榨不干。因为如果膏汁未尽的话,建茶的颜色就会显得浑浊,且味道过于浓重。"

"蒸压之后便是茶的干燥工序,叫作过黄。先是放在火上烘焙然后再放进沸水过,这样重复三次,再放在火上烘焙一夜,到了第二天,再过烟烘焙,火候不要过大,过大会导致茶饼面鼓泡,颜色焦黑。还注意防烟熏,熏过的茶会失去茶香而带焦糊味。因此以用温火为好。火力的强弱,要看銙的厚薄而定。如果是厚,那就过十到十五次的火;如果是薄銙,过六到八次,火候足了再过汤出色。出色之后,放进密室里用扇子紧扇,这样色泽就自然变得光洁莹亮。"

(姜怡、姜欣 译)

【译文】

Extracted from *Selected Information on Beiyuan Tea(Bei Yuan Bie Lu):*

"There are 46 imperial Tea Gardens, covering an area as spacious as 30 square *li*. Those listed ahead of Guanping are ranked as Neiyuan (Central Gardens), while those after Guankeng are ranked as Waiyuan (Marginal Gardens). When the trees begin to sprout early in spring, official tea gardens will bake green tea leaves a dozen days in advance of the civilian gardens. Some of the exceptionally good gardens like Jiuke (Nine Downfolds), Shi'erlong (Twelve Ridges), Longyou Ke (Roaming Dragon Downfold), Xiao Kuzhu (Small Bitter Bamboo), Zhang Keng (Open Hollow) and Xiji (West Side) will begin tea-processing ahead of other Imperial Gardens. Shimen (Stone Gate), Ruji (Lucky

Milk) and Xiangkou (Fragrant Mouth) are three private mills, often setting off their annual work five to seven days. When the tea is collected, steamed and pressed, it will be sent to Beiyuan for further processing.

"Tribute tea making mills there used to be divided into four sections. Problems arose due to the emulation and secret competition among the makers. Thus, the four sections are integrated into two, which explains the East Bureau and West Bureau in tea houses, and the East Brand and West Brand for the tea cakes. When the tea is taken out of the pestle basin, it should be blended thoroughly and kneaded smoothly. Then the paste is molded into cakes and baked. As for the molds for caked tea, there are fancy varieties, molds for large and small dragon balls. The molds have special names as labels in accordance with their ranks and batches to be sent to the imperial palace.

"Tea leaves should be plucked in the early morning before sunrise when the dews are not dried, which guarantee chubby and juicy leaves. Once the sun rises and shines, the moisture in the leaves will be sapped, and their quality will be affected. Consequently, when such sun stroke tea is boiled for drink, the color will not be as fresh and bright. So, tea collectors are always assembled at the Phoenix Mountain from 3 to 5 in the morning to pluck tea. (There is a pavilion just for the assembling purpose, and each time 220 laborers are needed.) An official supervisor will assign each worker a tablet. At about 7 o'clock , the striking sound of a *dong* will call back the laborers busy collecting in the mountain, in case they are too eager for more tea to keep the right timing. As a rule, the tea pickers all have a clear idea about the guidelines of tea collecting. Experienced natives and professionals are always preferred at recruiting, for they know well about everything concerned with the special task."

"It is necessary to differentiate the varieties of tea buds: small bud,medium bud, purple bud, whitish galeae, and dark pedicle. The small bud is shaped like the claw of hawks. The first batch of tribute made of 'small buds' includes Longtuan Shengxue (Snowy Dragon Ball) and White Tea, which are in turn steamed and plucked for the best in water.The leaves thin as needles are called 'water buds', making the top rank among 'small buds'. The medium bud is termed in ancient time as 'one spear (bud) and two flags (leaflets)'. The purple bud receives its name for the color of its leaves. Whitish galeae refer to the two leaflets encircling 'small buds'. And the dark pedicles are the leaflets at the end of each branch. Among all the varieties , 'Juicy Buds' rank the top, followed by 'small buds' and 'medium buds'. The rest is not even worth mentioning. Without careful selection, the tea will be stained a turbid color and have a pungent taste."

"Many plants on earth as well as insects will resuscitate from the Insects Waking Day. Tea should be plucked and processed three days before the day ,or after that date in case of a leap year when spring comes later than usual."

"Plucked tea buds should be cleaned thoroughly, put in a big steam box called zeng, and steamed on boiling water. However, steaming invites the dilemma: Being overdone, the buds will be stained with a yellowish color and lose some flavor. On the other hand, if underdone, tea buds will retain their greenish tang and raw grassy smell, and is liable to subside in cups. Therefore, a proper control over the steaming is a very vital procedure."

"The steamed tea should be washed and rinsed several times (to make it cold) before undergoing two other processes: slight pressing and strong pressing. The former is to remove the water , and the latter to eradicate the greasy juice (the delicate 'water buds' need yet another one, called heavy pressing), wrapped in cloth, and bound with bamboo peel. Tea makers then start to press. Up to midnight, the tea under constant pressing is taken out, kneaded evenly and sent to the second round of pressing. Tea makers have to work all night long hitting and pressing until the tea is deprived of every drop of liquid. This is because Jian'an Tea possesses a very thick and long-lasting taste, different from tea varieties from the lower region of the Yangtze River, where sap in tea is care-fully protected instead to reserve the precious taste. In the case of Jian'an Tea, however, the greasy sap has to be squeezed out entirely, as it may lead to turbid color and pungent taste."

"What follows is a series of drying procedures. The pressed tea is heated first by raging fire and then in boiling water. The two steps will be repeated a few times. The next morning, the tea needs yet another heating over a smoking fire. The fire should be smothered at a lukewarm degree, or the tea cakes may get burnt and blisters may appear, affecting the color and smoothness of the cakes' surface. Besides, too much smoke will deprive the tea of its fragrance and fine flavor. The level of attainment is up to the cake molds. If the cake is from a thick mold, stronger heatings are needed (about 10 to 15 times), and if it is of a thin mold, a mild heating is necessary (6 to 8 times). After the firing, the tea is to be boiled again for a better color. When it is done, the workers will just fan the tea swiftly in close tea rooms as a final touch. With all these procedures finished, the tea cakes will obtain a natural lustrous tint."

(姜怡、姜欣 译)[①]

① 陆羽，陆廷灿. The Classic of Tea. The Sequel to The Classic of Tea Vol. Ⅰ, Ⅱ. 茶经. 续茶经(全二册)[M]. 姜怡，姜欣译. 长沙：湖南人民出版社，2019.

3.[先秦]《甫田》——佚名

无田甫田,维莠骄骄。无思远人,劳心忉忉。

无田甫田,维莠桀桀。无思远人,劳心怛怛。

婉兮娈兮。总角丱兮。未几见兮,突而弁兮!

【原文解读】

大田宽广不可耕,野草高高长势旺。切莫挂念远方人,惆怅不安心惶惶。大田宽广不可耕,野草深深长势强。切莫挂念远方人,惆怅不安心怏怏。漂亮孩子逗人怜,扎着小小羊角辫。才只几天没见面,忽戴冠帽已成年。

【译文一】

FU TIAN

Do not try to cultivate fields too large;
—The weeds will only grow luxuriantly.
Do not think of winning people far away;
—Your toiling heart will be grieved.
Do not try to cultivate fields too large;
—The weeds will only grow proudly .
Do not think of winning people far away ;
—Your toiling heart will be distressed.
How young and tender,
ls the child with his two tufts of hair!
When you see him after not a long time,
Lo ! he is wearing the cap ![1]

(James 译)

① University of Virgina Library

【译文二】

甫　田

无田甫田，　　　　主子大田别去种，
维莠骄骄。　　　　野草茂盛一丛丛。
无思远人，　　　　远方人儿别想他，
劳心忉忉。　　　　见不到他心伤痛。

无田甫田，　　　　主子大田别去耪，
维莠桀桀。　　　　野草长得那么旺。
无思远人，　　　　远方人儿别想他，
劳心怛怛。　　　　见不到他徒忧伤。

婉兮娈兮，　　　　小小年纪多姣好，
总角丱兮。　　　　两束头发象羊角。
未几见兮，　　　　不久倘能见到他，
突而弁兮？　　　　突然戴上成人帽。

（程俊英 译诗）

MISSING HERSON*

Don't till too large a ground,
Or weed will spread around.
Don't miss one far away,
Or you'll grieve night and day.

Don't till too large a ground,
Or weed overgrows around.
Don't miss the far-off one,

Or your grief won't be done.

My son was young and fair,
With his two tufts of hair.
Not seen for a short time,
He's grown up to his prime.

*It was said that this song was written for Wen Jiang of Qi (See the preceding poem) missing her son who became Duke Zhuang of Lu at the age of thirteen.

(许渊冲 译)①

① 佚名.Book of Poetry诗经[M].许渊冲,译.北京:五洲传播出版社,2011.

第五章　农田技术篇

◎**本章学习目标**◎

1. 农田技术在农业科技典籍中的概述
2. 赏析农田技术的典籍双语佳作

第一节　农田技术在典籍中的概述

我国古代的农业科学，是在封建社会小农经济的基础发展起来的，具有其历史的局限性。但它的成就同欧洲古代和中世纪的农业科学相比，确实是独具一格、光彩夺目的。十九世纪欧洲杰出的农业化学家李比希说过：中国的农业“是以经验和观察为指导，长远地保持着土壤肥力，借以适应人口的增长而不断提高其产量，创造了无与伦比的农业耕种方法”[①]。《尚书》中记载了古代农业的各个方面，包括耕种、播种、施肥、灌溉等。例如，在《尚书·洪范》中，记载了古代农业的基本原则，强调了农民应该根据季节的变化来选择合适的农事活动，以确保农作物的生长和收获。此外，在《尚书·周书》

① 华北农业大学农业科学技术史研究组. 精耕细作——我国古代农业科学技术的优良传统（一）[J]. 中国农业科学，1978，11（1）：91-96.

中,还有关于土地利用、农田水利等方面的记载,这些内容对于我们了解古代农业的发展和演变具有重要意义。[①]在《尚书·禹贡》中,记载了古代政府对农业生产的监督和管理,通过对土地的分配和税收的征收,来保障农民的生产积极性和国家的粮食供应。

农田水利之始,大约可以追溯到先周时期。与后稷同时的夏禹决江疏河,致力乎沟洫事业,并且田彼南山,导引注水,使之东流。周祖后稷列封疆,画畔界,辨水土之所宜。后稷之侄,名曰叔均,他建议置旱神于赤水以北,除水道通沟渎以避旱灾,则第一次将水利措施与农业生产联系在了一起。西周关中地区既有排除湿地积水的沟洫之制,亦有蓄水引灌的小型工程。《诗·小雅·白桦》中所言的滮池灌溉工程,应是我国见诸文字记载的第一个农田水利工程。战国末期,秦修郑国渠,陕西始有大型引河凿渠灌溉工程。汉魏间,陕西水利建设形成高潮,关中四大水系(泾、驸、洛、汧)全都用诸农田水利,关中水利网初现雏形,后世渠系基本缘此。[②]

《陈旉农书》是最早一部介绍中国江南水稻地区栽培技术的典型地方性农书。烤田是水稻灌溉制度中的重要内容;是在水稻生长某些阶段,利用放干田中水层,使土壤曝晒的过程。烤田技术首先见于北魏著名农学家贾思勰的《齐民要术》。他说:三日之中,令人驱鸟。稻苗长七八寸,陈草复起,以镰侵水芟之,草悉脓死。稻苗渐长,复须薅。拔草曰"薅"。虎高反。薅讫,决去水,曝根令坚。量时水旱而溉之。将熟,又去水。霜降获之。[③]

在稻田耕地之后,"随于中间及四傍为深大之沟,俾水竭涸,泥坼裂而极干,然后作起沟缺,次第灌溉。夫已干燥之泥,骤得雨即苏碎,不三五日间,稻苗蔚然,殊胜于用粪也"[④]。烤田除了有促进水稻根系发育的作用,还可以

① 百度文库.

② 樊志民.陕西古代农田水利科学技术初探[J].水资源与水工程学报,1990(3):84-91.

③ 北魏·贾思勰《齐民要术》,万有文库本,第22页.

④ 陈旉.陈旉农书校释[M].刘铭,校释.北京:中国农业出版社,1956.

提高土壤温度，改善土壤透气性，有利于好气性微生物活动，加速土壤有机物的分解，起到“殊胜用粪”的增肥效果，达到促进稻株分蘖（“稻苗蔚然”）的目的。南宋年间，烤田技术已在江浙一带普及。高斯得在出任宁国府（治今安徽宣城）太守时曾将这一技术在当地加以推广。”

最早记载灌溉技术的典籍见于西汉末年成书的《氾胜之书》。他在论述稻田灌溉时说：“始种，稻欲温。温者，缺其塍，令水道相直；夏至后，太热，令水道错。”[①]“灌溉水源温度一般低于田中土壤温度，因此当水稻初种时，为了尽可能保持地温，就要在由一块田向下一块田灌水时，将上一田块进出水口笔直相对，这样，灌溉水可径直通过，对上一田块原有的较高水温的水层扰动较少，地温得以保持；而在高温季节，需要用灌溉水降低土壤温度时，则将田块进出水口错开，流水斜穿田块，新水较多地代替田内存水，有助于降低田块土壤温度。”[②]

农田技术在水利工程、种植技术与养殖技术等方面都在古代典籍里有记载。例如，《山海经》中记载了大禹治水的故事，描述了利用河道调节水流的方法。《淮南子》中记载了修筑引水渠的技术，使得农田得以灌溉。《周礼》中还提到了种植不同作物的时机和方法。《礼记·檀弓上》中提到了养蚕的方法，描述了蚕的饲养和蚕茧的采摘。《尔雅》中还有关于养鸡、养猪等畜牧技术的记载。

---

① 万国鼎.氾胜之书辑释[M].北京：农业出版社，1980：121.

② 周魁一.中国古代农田灌溉排水技术[J].古今农业，1997，121(1)：1-12.

第二节　农田技术典籍译作赏析

【赏析一】

《管子》地员　第五十八

夫管仲之匡天下也，其施七尺。

渎田悉徙，五种无不宜，其立后而手实。其木宜蚖、蓄与杜、松，其草宜楚棘。见是土也，命之曰五施，五七三十五尺而至于泉。呼音中角。其水仓，其民强。

赤垆，历强肥，五种无不宜。其麻白，其布黄，其草宜白茅与雚，其木宜赤棠。见是土也，命之曰四施，四七二十八尺而至于泉。呼音中商。其水白而甘，其民寿。

黄唐，无宜也，唯宜黍秫也。宜县泽。行廧落，地润数毁，难以立邑置廧。其草宜黍秫与茅，其木宜櫄、桑。见是土也，命之曰三施，三七二十一尺而至于泉。呼音中宫。其泉黄而糗，流徙。

斥埴，宜大菽与麦。其草宜萯、雚，其木宜杞。见是土也，命之曰再施，二七一十四尺而至于泉。呼音中羽。其泉咸，水流徙。

黑埴，宜稻麦。其草宜苹、蓨，其木宜白棠。见是土也，命之曰一施，七尺而至于泉。呼音中徵。其水黑而苦。

凡听徵，如负猪豕觉而骇。凡听羽，如鸣马在野。凡听宫，如牛鸣窌中。凡听商，如离群羊。凡听角，如雉登木以鸣，音疾以清。凡将起五音凡首，先主一而三之，四开以合九九，以是生黄钟小素之首，以成宫。三分而益之以一，为百有八，为徵。不无有三分而去其乘，适足，以是生商。有三分，而复于其所，以是成羽。有三分，去其乘，适足，以是成角。

坟延者，六施，六七四十二尺而至于泉。陕之芳，七施，七七四十九尺而至于泉。祀陕八施，七八五十六尺而至于泉。杜陵九施，七九六十三尺而至于泉。延陵十施，七十尺而至于泉。环陵十一施，七十七尺而至于泉。蔓山十二施，八十四尺而至于泉。付山十三施，九十一尺而至于泉。付山白徒十四施，九十八尺而至于泉。中陵十五施，百五尺而至于泉。青山十六施，百一十二尺而至于泉，青龙之所居，庚泥，不可得泉，赤壤𫵧山十七施，百一十九尺而至于泉，其下清商，不可得泉。𨺅山白壤十八施，百二十六尺而至于泉，其下骈石，不可得泉。徙山十九施，百三十三尺而至于泉，其下有灰壤，不可得泉。高陵土山二十施，百四十尺而至于泉。

山之上，命之曰县泉，其地不干，其草如茅与莼，其木乃樠，凿之二尺，乃至于泉。山之上，命之曰复吕，其草鱼肠与莸，其木乃柳，凿之三尺而至于泉。山之上，命之曰泉英，其草蕲、白昌，其木乃杨，凿之五尺而至于泉。山之材，其草兢与蔷，其木乃格，凿之二七十四尺而至于泉。山之侧，其草葍与蒌，其木乃品榆，凿之三七二十一尺而至于泉。

凡草土之道，各有谷造。或高或下，各有草土。叶下于𩇯，𩇯下于苋，苋下于蒲，蒲下于苇，苇下于雚，雚下于蒌，蒌下于荓，荓下于萧，萧下于薜，薜下于萑，萑下于茅。凡彼草物，有十二衰，各有所归。

九州之土，为九十物。每州有常，而物有次。

群土之长，是唯五粟。五粟之物，或赤，或青，或白，或黑，或黄，五粟五章。五粟之状，淖而不韧，刚而不觳，不泞车轮，不污手足。其种，大重、细重，白茎、白秀，无不宜也。五粟之土，若在陵在山，在隫在衍，其阴其阳，尽宜桐柞，莫不秀长。其榆其柳，其檿其桑，其柘其栎，其槐其杨，群木蕃滋，数大条直以长。其泽则多鱼，牧则宜牛羊。其地其樊，俱宜竹、箭、藻、龟、楢、檀。五臭生之：薜荔、白芷，麋芜、椒、连。五臭所校，寡疾难老，士女皆好，其民工巧。其泉黄白，其人夷姤。五粟之土，干而不格，湛而不泽，无高下，葆泽以处。是谓粟土。

粟土之次曰五沃。五沃之物,或赤,或青,或黄,或白,或黑。五沃五物,各有异则。五沃之状,剽忞橐土,虫易全处,忞剽不白,下乃以泽。其种,大苗细苗,赨茎黑秀箭长。五沃之土,若在丘在山,在陵在冈,若在陬,陵之阳,其左其右,宜彼群木,桐、柞、枎、櫄,及彼白梓。其梅其杏,其桃其李,其秀生茎起,其棘其棠,其槐其杨,其榆其桑,其杞其枋,群木数大,条直以长。其阴则生之楂藜,其阳则安树之五麻,若高若下,不择畴所。其麻大者如箭,如苇,大长以美;其细者如雚,如蒸。欲有与各,大者不类,小者则治,揣而藏之,若众练丝,五臭畴生,莲、与、藁芜,藁本白芷。其泽则多鱼,牧则宜牛羊。其泉白青;其人坚劲,寡有疥骚,终无痟酲。五沃之土,干而不斥,湛而不泽,无高下,葆泽以处。是谓沃土。

Categories of Land

When Guan Zhong was governing the state of Qi, seven *chi* were stipulated to be one *shi*.

Fertile lands along riversides are suitable to grow all kinds of crops. Crops growing at such places will develop big seeds and heavy ears. Trees suitable to grow there are Yuan, Cang, birch-leaf pears and pines. Thorns can grow well there as well. This kind of soil is addressed as Five *Shi*. Five times seven, so, the groundwater is thirty-five *chi* beneath the ground surface. When people shout at the top of their voices, it sounds somewhat the same as the note of Jue. Water there is dark green. People living there are robust.

The Chi Lu soil is dry, loose, hard and fertile, and it is suitable for all kinds of crops. Hemp produced there is white and the fabric made of it is yellow. Grasses growing well there are cogon and reeds. Trees growing well there are Chi Tang. This kind of soil is addressed as Four *Shi*. Four times seven, so the groundwater is twenty-eight *chi* beneath the ground surface. When people shout at the top of their voices, it sounds somewhat the same as the note of Shang. Water there is white and sweet. People living there have long lifespans.

Normally, the Huang Tang soil is not suitable for growing crops. Only millet and broom corn can grow well there. Water should be drained from this kind of soil. When walls are built with this kind of earth, they often collapse while the earth is moist. Accordingly, it is very difficult to erect towns and walls there. Grasses growing well there are thistles and thatch grass. Trees *growing* well there are the long-living trees, all-age trees and mulberry trees. This kind of soil is addressed as Three *Shi*. Three times seven, so the groundwater is twenty-one chi beneath the ground surface. When people shout at the top of their voices, it sounds somewhat the same as the note of Gong. Water there is yellow, stinky and runs off easily.

The Chi Zhi soil is suitable to grow beans and barley. Grasses growing well in that area are Fu and Guan.Chinese wolfberries can grow well there. This kind of soil is addressed as Zai (two) *Shi*. Two times seven, so that groundwater is fourteen *chi* beneath the ground surface. When people shout at the top of their voices, it sounds somewhat the same with the note of Yu. Water there is salty and runs off quickly.

The Hei Zhi soil is suitable to grow rice and barley. Grasses growing well on that kind of soil are wormwood and the "sheep-hoof grass". Birch-leaf trees can grow well there. This kind of soil is addressed as One *Shi*. The groundwater is seven *chi* beneath the ground surface. When people shout at the top of their voices, it sounds somewhat the same with the note of Zhi. Water there is black and bitter.

The note of Zhi sounds like a sow roaring when it witnesses her piglet being carried away. The note of Yu sounds like a horse neighing in the wild. The note of Gong sounds like a cow bellowing in the cellar. The note of Shang sounds like a sheep crying when it is separated from the herd. The note of Jue sounds rapid and clear, just like a pheasant singing in the woods. The five notes should be established in this way: set up a string of fixed length and divide it into three parts equally. And then divide each part into three smaller parts and repeat that two times again, the result would be eighty-one parts. This pitch would be regarded as Huang Zhong Xiao Su. This is how Gong is determined. Add one third of the fixed length to the string and then divide it into four parts equally; after that divide each of these four parts three times by three, and the result should be one hundred and eight parts. This is how Zhi is determined. Divide the above-mentioned result by three and then subtract the remainder; the result is regarded as Shang. Divide the above-mentioned result by three and then add the remainder; in this way Yu is determined. Divide the above-mentioned result by three and then subtract the remainder; in this way Jue is determined.

Fen Yan (referring to sloping fields which are relative high) is six *Shi*. Six times seven, so the groundwater is forty-two *chi* beneath the ground surface.Shan Zhi Fang (referring to the land along both sides of narrow valleys) is seven *Shi*. Seven times seven, so the groundwater is forty-nine *chi* beneath the ground surface. Si Shan (referring to the land around valleys) is eight *Shi*. Eight times seven, so the ground water is fifty-six *chi* beneath the ground surface. Du Ling (referring to the hills) is nine *Shi*. Nine times seven,so,the groundwater is sixty-three *chi* beneath the ground surface. Yan Ling (referring to areas around the hills) is ten *Shi*. The groundwater is seventy *chi* beneath the ground surface.

Huan Ling (referring to areas around mountains) is eleven *Shi*. The groundwater is seventy-seven *chi* beneath the ground surface. Man Shan (referring to mountain ranges) is twelve *Shi*. The groundwater is eighty - four *chi* beneath the ground surface. Fu Shan (referring to areas around mountains) is thirteen *Shi*. The groundwater is ninety - one *chi* beneath the ground surface. Fu Shan Bai Tu (referring to areas around mountains, where the earth is white) is fourteen *Shi*. The groundwater is ninety-eight *chi* beneath the ground surface. Zhong Ling (referring to the centers of hills) is fifteen *Shi*. The groundwater is one hundred and five *chi* beneath the ground surface. Qing Shan (referring to green mountains) is sixteen *Shi*. The groundwater is one hundred and twelve *chi* beneath the ground surface. At the place where the black dragon lives, the soil is rigid and there are no underground springs. Chi Rang Ao Shan (referring to small mountains with red earth) is seventeen *Shi*. The ground water is one hundred and nineteen *chi* beneath the ground surface. Qing Shang stays beneath there, and there are no underground springs. Cuo Shan Bai Rang (referring to mountainous areas with white earth) is eighteen *Shi*. The groundwater is one hundred and twenty six *chi* beneath the ground surface. A lot of stones are beneath it, but there are no underground springs. Dou Shan (referring to steep mountains) is nineteen *Shi*. The groundwater is one hundred and thirty three chi beneath the ground surface. The soil beneath it is gray, and there are no underground springs. Gao Ling Tu Shan (referring to earth mountains located on lands of high altitude) is twenty *Shi*. The groundwater is one hundred and forty *chi* beneath the ground surface.

On tops of mountains can be places known as Xuan Quan. Soil there will never dry up. Grasses growing there are Indian madder and thatch. Trees growing there are Man. Digging two *chi* deep in the soil can reach underground water. On tops of mountains can be places known as Fu Lu. Grasses growing there are "fish-gut grass" and bluebeards. Willows grow there as well. Digging three *chi* deep in the soil can reach underground water. On tops of mountains can be places known as Quan Ying. Grasses growing there are Chinese angelicas and sweet flags. Aspen trees grow there too. Digging five *chi* deep in the soil can reach underground water. On mountainsides grasses such as Saint Paul 's wort and confervoides grow. Chinese catalpa trees grow there too. Two times seven, so digging fourteen *chi* deep in the soil can reach underground water. On mountainsides grasses such as Bi and beach worm woods grow. Bristle elms grow there too. Digging twenty-one *chi* deep in the soil can reach underground water.

Certain kinds of plants should grow on certain kinds of soil. There are grasses growing everywhere, no matter whether it is high or low-lying. Seaweed only with leaves grows at a height lower than water chestnut. Water chestnut grows at a height lower than three-coloured amaranth. Three-coloured amaranth grows at a height lower than sweet flag. Sweet flag grows at a height lower than reed. Reed grows at a height lower than the Guan. Guan grows at a height lower than beach wormwood. Beach wormwood grows at a height lower than Bing. Bing grows at a height lower than wormwood. Wormwood grows at a height lower than creeping fig. Creeping fig grows at a height lower than motherwort. Motherwort grows at a height lower than thatch grass.There are twelve kinds of grasses, and they all grow in certain areas.

There are ninety kinds of soil among all the nine *zhou*. And the kinds of soil in each *zhou* have inherent characteristics that make them suitable for specific corresponding plants.

The five kinds of Su Tu are of the highest quality. Regarding the colours of these five Su Tu: some are red,some are green, some are white, some are black, and some are yellow. These five colours are the five signs of Su Tu which have inherent characteristics: some are moist but not sticky, some are dry but not sterile and both do not adhere to wheels or cling to hands and feet. Crops can grow well in them, whether they are inseminated sparsely or densely. If they have white stems and white ears. These five kinds of soil, whether they occur in hilly or mountainous areas, near waters or on plains, north facing or south facing, are suitable for phoenix trees and oaks that can grow very tall. Other trees such as elms, willows, Yan, mulberries, silkworm thorns, saw-tooth oaks, Japanese pagoda trees and aspens can all grow well and fast, their crowns are huge and their trunks are straight and tall. There are a lot fish in the waters. And these kinds of soil also provide excellent meadow for cows and sheep. The fields and mountainsides with these kinds of soil are suitable for bamboos, China-canes, Chinese dates, teaks, sandal woods, etc. Five fragrant plants grow there: creeping figs, angelicas, confervoides, Bunge prickly ashes and eupatorium. They can make people healthy and stay youthful. They can make men handsome, make women pretty, and make everyone clever and deft. The water there is yellow white. The countenances of people living there are fine and smooth. These five kinds of Su Tu are dry but not hard, moist but not loose. They can keep soil moisture no matter whether they are high or low-lying. These kinds of soil are called Su Tu.

Soils inferior to Su Tu are the five kinds of Wo Tu. Regarding these five kinds of Wo Tu: some are red, some are green, some are white, some are black and some are yellow. They have five different colours, so they can be differentiated from one another. They also have their own inherent characteristics: they are loose and full of pores, so that many insects and worms live there. They are hard but do not look white on the surface when they are dry, and they hold moisture beneath the surface. Crops can grow well whether they are inseminated sparsely or densely, and they will develop red stems and black ears with long awns. As for these five Wo Tu, whether they are located among hills, mountains, mounds or ridges, or whether they are just narrow margins or on the southern side of mounds, they are suitable to plant phoenix trees, oaks, red hibiscuses, long-living trees and Chinese catalpas. Japanese apricots, almonds, peaches and plums can grow tall and develop a lot of flowers, trees. Thorns, birch-leaf pera trees, aspens, elms, mulberry trees, Chinese wolfberry trees and sandal woods all grow quickly with long, straight branches and high trunks. Hawthorns and pears can grow in places not exposed to the sun in these area. And all kinds of hemp plants can be planted everywhere at places exposed to the sun, no matter whether these places are high or low-lying. Big hemp plants like China-canes and reeds grow tall and are of good quality. Small ones like slight firewoods are numerous and should be planted in orderly reeds and rows. The big ones do not have gnarls and the small ones are easy to cut. When they are rolled into balls and kept for latter use, they become very white, like bleached silk. Five fragrant plants grow there: orchids, Asiatic plantains, confervoides, angelicas, etc. There are a lot of fish in the confervoides. And these kinds of soil also provide excellent meadows for cows and sheep. The water there is light green. People living there are robust and seldom suffer from diseases such as scabies, headache or vertigo. These five kinds of Wo Tu are dry but not desiccated, moist but not loose. They can keep soil moisture no matter whether they are high or low-lying. These kinds of soil are called Wo Tu.[①]

(翟江月 译)

① 管仲. Guanzi Vol. Ⅰ, Ⅱ, Ⅲ, Ⅳ 管子(全四册)[M]. 翟江月, 译. 桂林:广西师范大学出版社, 2005.

【赏析二】

《尚书》节选

注:《洪范》记载了古代农业的基本原则,强调了农民应该根据季节的变化来选择合适的农事活动,以确保农作物的生长和收获。

惟十有三祀,王访于箕子。王乃言曰:"呜呼!箕子,惟天阴骘下民,相协厥居,我不知其彝伦攸叙。"

箕子乃言曰:"我闻在昔,鲧堙洪水,汩陈其五行。帝乃震怒,不畀洪范九畴,彝伦攸斁。鲧则殛死,禹乃嗣兴,天乃锡禹洪范九畴,彝伦攸叙。"

初一曰五行,次二曰敬用五事,次三曰农用八政,次四曰协用五纪,次五曰建用皇极,次六曰乂用三德,次七曰明用稽疑,次八曰念用庶征,次九曰向用五福,威用六极。

一、五行:一曰水,二曰火,三曰木,四曰金,五曰土。水曰润下,火曰炎上,木曰曲直,金曰从革,土爰稼穑。润下作咸,炎上作苦,曲直作酸,从革作辛,稼穑作甘。

二、五事:一曰貌,二曰言,三曰视,四曰听,五曰思。貌曰恭,言曰从,视曰明,听曰聪,思曰睿。恭作肃,从作乂,明作哲,聪作谋,睿作圣。

三、八政:一曰食,二曰货,三曰祀,四曰司空,五曰司徒,六曰司寇,七曰宾,八曰师。

四、五纪:一曰岁,二曰月,三曰日,四曰星辰,五曰历数。

五、皇极:皇建其有极。敛时五福,用敷锡厥庶民,惟时厥庶民于汝极。锡汝保极:凡厥庶民,无有淫朋,人无有比德,惟皇作极。凡厥庶民,有猷有为有守,汝则念之。不协于极,不罹于咎,皇则受之。而康而色,曰:'予攸好德。'汝则锡之福。时人斯其惟皇之极。无虐茕独而畏高明,人之有能有为,使羞其行,而邦其昌。凡厥正人,既富方榖,汝弗能使有好于而家,时人斯其辜。于其无好德,汝虽锡之福,其作汝用咎。无偏无陂,遵王之义;无有作好,遵王之道;无有作恶,遵王之路;无偏无党,王道荡荡;无党无偏,王道平平;无反无侧,王道正直。会其有极,归其有极。曰:皇,极之敷言,是彝是训,于帝其训。凡厥庶民,极之敷言,是训是行,以近天子之光。曰:天子作民父母,以为天下王。

六、三德:一曰正直,二曰刚克,三曰柔克。平康、正直;强弗友、刚克;燮友,柔克。沉潜,刚克;高明,柔克。惟辟作福,惟辟作威,惟辟玉食。臣无有作福作威玉食。臣之有作福作威玉食,其害于而家,凶于而国。人用侧颇僻,民用僭忒。

七、稽疑:择建立卜筮人,乃命卜筮。曰雨,曰霁,曰蒙,曰驿,曰克,曰贞,曰悔,凡七。卜五,占用二,衍忒。立时人作卜筮。三人占,则从二人之言。汝则有大疑,谋及乃心,谋及卿士,谋及庶人,谋及卜筮。汝则从,龟从,筮从,卿士从,庶民从,是之谓大同。身其康强,子孙其逢。吉。汝则从,龟从,筮从,卿士逆,庶民逆,吉。卿士从,龟从,筮从,汝则逆,庶民逆,吉。庶民从,龟从,筮从,汝则逆,卿士逆,吉。汝则从,龟从,筮逆,卿士逆,庶民逆,作内吉,作外凶。龟筮共违于人,用静吉,用作凶。

八、庶征:曰雨,曰旸,曰燠,曰寒,曰风。曰时五者来备,各以其叙,庶草蕃庑。一极备,凶;一极无,凶。曰休征:曰肃,时雨若;曰乂,时旸若;曰晢,时燠若;曰谋,时寒若;曰圣,时风若。曰咎征:曰狂,恒雨若;曰僭,恒旸若;曰豫,恒燠若;曰急,恒寒若;曰蒙,恒风若。曰王省惟岁,卿士惟月,师尹惟日。岁月日时无易,百谷用成,乂用明,俊民用章,家用平康。日月岁时既易,百谷用不成,乂用昏不明,俊民用微,家用不宁。庶民惟星,星有好风,星有好雨。日月之行,则有冬有夏。月之从星,则以风雨。

九、五福:一曰寿,二曰富,三曰康宁,四曰攸好德,五曰考终命。六极:一曰凶、短、折,二曰疾,三曰忧,四曰贫,五曰恶,六曰弱。

The Great Plan

In the thirteenth year, the king went to enquire of the count of Qi, and said to him, "Oh! count of Qi, Heaven, (working) unseen, secures the tranquility of the lower people, aiding them to be in harmony with their condition①. I do not know how the unvarying principles (of its method in doing so) should be set forth in due order."

The count of Qi thereupon replied, "I have heard that in old time Gun dammed up the inundating waters, and thereby threw into disorder the arrangement of the five elements. God was consequently roused to anger, and did not give him the Great Plan with its nine divisions, and thus the unvarying principles (of Heaven's method) were allowed to go to ruin. Gun was therefore kept a prisoner till his death, and his son Yü rose up (and entered on the same undertaking). To him Heaven gave the Great Plan with its nine divisions, and the unvarying principles (of Heaven's method) were set forth in their due order."

(Of those divisions) the first is called "the five elements"; the second, "reverent attention to the five (personal) matters"; the third, "earnest devotion to the eight (objects of) government"; the fourth, "the harmonious use of the five dividers of time"; the fifth, "the establishment and use of royal perfection"; the sixth, "the discriminating use of the three virtues"; the seventh, "the intelligent use of (the means for) the examination of doubts"; the eighth, "the thoughtful use

---

① Kong Ying-da of the Tang Dynasty says on this: —"The people have been produced by supreme Heaven, and both body and soul are Heaven's gift. Men have thus the material body and the knowing mind, and Heaven further assists them, helping them to harmonize their lives. The right and the wrong of their language, the correctness and errors of their conduct, their enjoyment of clothing and food , the rightness of their various movements; —all these things are to be harmonized by what they are endowed with by Heaven."

of the various verifications"; the ninth, "the hortatory use of the five (sources of) happiness, and the awing use of the six (occasions of) Suffering."

i. First, of the five elements. —The first is water; the second is fire; the third, wood; the fourth, metal; and the fifth, earth. (The nature of) water is to soak and descend; of fire, to blaze and ascend; of wood, to be crooked and straight; of metal, to yield and change; while (that of) earth is seen in seed-sowing and in-gathering. That which soaks and descends becomes salt; that which blazes and ascends becomes bitter; that which is crooked and straight becomes sour; that which yields and changes becomes acrid; and from seed-sowing and in-gathering comes sweetness.

ii. Second, of the five (personal) matters. —The first is the bodily demeanour; the second, speech; the third, seeing; the fourth, hearing; the fifth, thinking. (The virtue of) the bodily appearance is respectfulness; of speech, accordance (with reason); of seeing, clearness; of hearing, distinctness; of thinking, perspicaciousness. The respectfulness becomes manifest in gravity; accordance (with reason), in orderliness; the clearness, in wisdom; the distinctness, in deliberation; and the perspicaciousness, in sageness.

iii. Third, of the eight (objects of) government. —The first is food; the second, wealth and articles of convenience; the third, sacrifices; the fourth, (the business of) the Minister of Works; the fifth, (that of) the Minister of Instruction; the sixth, (that of) the Minister of Crime; the seventh, the observances, to be paid to guests; the eighth , the army.

iv. Fourth, of the five dividers of time. —The first is the year (or the planet Jupiter); the second, the moon; the third, the sun; the fourth, the stars and planets, and the zodiacal spaces; and the fifth, the calendaric calculations.

v. Fifth, of royal perfection. —The sovereign, having established (in himself) the highest degree and pattern of excellence, concentrates in his own person the five (sources of) happiness, and proceeds to diffuse them, and give them to the multitudes of the people. Then they, on their part, embodying your perfection, will give it (back) to you, and secure the preservation of it. Among all the multitudes of the people there will be no unlawful confederacies, and among men (in office) there will be no bad and selfish combinations; —let the sovereign establish in (himself) the highest degree and pattern of excellence. Among all the multitudes of the people there will be those who have ability to plan and to act, and who keep themselves (from evil): —do you keep such in mind; and there will be those who , not coming up to the highest point of excellence, yet do not involve themselves in evil: —let the sovereign receive such. And when a placid satisfaction appears in their countenances , and they say, 'Our love is fixed on virtue', do you then confer favours on them; —those men will in this way advance to the perfection of the sovereign. Do not let him oppress the friendless and childless, nor let him fear the high and distinguished. When men (in office) have ability and administrative power, let them be made still more to cultivate their conduct; and the prosperity of the country will be promoted. All (such) right men, having a competency, will go on in goodness. If you cannot cause them to have what they love in their families, they will forthwith proceed to be guilty of crime. As to those who have not the love of virtue, although you confer favours (and emoluments) on them, they will (only) involve you in employing the evil.

"Without deflection, without unevenness,

Pursue the royal righteousness.

Without selfish likings,

Pursue the royal way.

Without selfish dislikings ,

Pursue the royal path.

Avoid deflection, avoid partiality ;

Broad and long is the royal way.

Avoid partiality, avoid deflection;

Level and easy is the royal way.

Avoid perversity, avoid one sidedness;

Correct and straight is the royal way.

(Ever) seek for this perfect excellence,

(Ever) turn to this perfect excellence."

He went on to say, "This amplification of the royal perfection contains the unchanging (rule), and is the (great) lesson; —yea, it is the lesson of God. All the multitudes of the people, instructed in this amplification of the perfect excellence, and carrying it into practice, will thereby approximate to the glory of the Son of Heaven." And say, "The Son of Heaven is the parent of the people, and so becomes the sovereign of all under the sky."

vi. Sixth, of the three virtues. —The first is correctness and straightforwardness; the second, strong rule; and the third, mild rule. In peace and tranquillity, correctness and straightforwardness (must sway); in violence and disorder, strong rule; in harmony and order, mild rule. For the reserved and retiring there should be (the stimulus of) the strong rule; for the high (-minded) and distinguished, (the restraint of) the mild rule. It belongs only to the sovereign to confer dignities and rewards, to display the terrors of majesty, and to receive the revenues (of the kingdom). There should be no such thing as a minister's conferring dignities or rewards, displaying the terrors of majesty, or receiving the revenues. Such a thing is injurious to the clans, and fatal to the states (of the kingdom); smaller affairs are thereby managed in a one-sided and perverse manner, and the people fall into assumptions and excesses.

vii. Seventh, of the (means for the) examination of doubts. —Officers having been chosen and appointed for divining by the tortoise-shell and the stalks of the Achillea, they are to be charged (on occasion) to execute their duties. (In doing this), they will find (the appearances of) rain, of clearing up, of cloudiness, of want of connexion, and of crossing; and the inner and outer diagrams. In all (the indications) are seven; —five given by the shell, and two by the stalks; and (by means) of these any errors (in the mind) may be traced out. These officers having been appointed, when the divination is proceeded with, three men are to interpret the indications, and the (consenting) words of two of the mare to be followed. When you have doubts about any great matter, consult with your own mind; consult with your high ministers and officers; consult with the common people; consult the tortoise-shell and divining stalks. If you, the shell, the stalks, the ministers and officers,and the common people, all agree about a course, this is what is called a great concord, and the result will be the welfare of your person and good fortune to your descendants. If you, the shell, and the stalks agree, while the ministers, and officers,with the shell and stalks, agree, while you and the common people oppose, the result will be fortunate. If the common people, the shell, and the stalks agree, while you, with the ministers and officers, oppose, the result will be fortunate. If you and the shell agree, while the stalks, with the ministers and officers, and the common people oppose, internal operations will be fortunate, and external undertakings unlucky. When the shell and talks are both opposed to the views of men, there will be good fortune in being still, and active operations will be unlucky.

viii. Eighth, of the various verifications. —They are rain, sunshine, heat, cold, wind, and seasonableness. When the five come, all complete, and each in its proper order, (even) the various plants will be richly luxuriant. Should any one of them be either excessively abundant or excessively deficient, there will be evil. There are the favourable verifications: —namely, of gravity, which is emblemed by seasonable rain; of orderliness, emblemed by seasonable sunshine;of wisdom, emblemed by seasonable heat; of deliberation, emblemed by seasonable cold; and of sageness, emblemed by seasonable wind. There are (also) the unfavourable verifications: —namely, of recklessness, emblemed by constantrain; of assumption, emblemed by constant sunshine; of indolence, emblemed by constant heat; of hastiness, emblemed by constant cold; and of stupidity, emblemed by constant wind.

He went on to say, "The king should examine the (character of the whole) year; the high ministers and officers (that of) the month; and the inferior officers (that of) the day. If, throughout the year, the month, the day, there be an unchanging seasonableness, all the grains will be matured; the measures of government will be wise, heroic men will stand forth distinguished; and in the families (of the people) there will be peace and prosperity. If, throughout the year, the month, the day, the season ableness be interrupted, the various kinds of grain will not be matured; the measures of government will be dark and unwise; heroic men will be kept in obscurity; and in the families (of the people) there will be an absence of repose. By the common people the stars should be examined. Some stars love wind, and some love rain. The courses of the sun and moon give winter and summer. The way in which the moon follows the stars gives wind and rain."

ix. Ninth, of the five (sources of) happiness. —The first is long life; the second, riches; the third, soundness of body and serenity of mind; the fourth, the love of virtue; and the fifth, fulfilling to the end the will (of Heaven). Of the six extreme evils, the first is misfortune shortening the life; the second, sickness; the third, distress of mind; the fourth, poverty; the fifth, wickedness; the sixth, weakness.

(理雅各 英译)

注:在《尚书·周书》中,有关于农田改良的记载,提到了古代农民通过开垦荒地、疏浚河渠等方式来改良土地质量。这些记载不仅有助于我们了解古代农业技术的发展,还对于今天的农业生产具有一定的借鉴意义。

吕 刑

惟吕命,王享国百年,耄,荒度作刑,以诘四方。

王曰:"若古有训,蚩尤惟始作乱,延及于平民,罔不寇贼,鸱义,奸宄,夺攘、矫虔。苗民弗用灵,制以刑,惟作五虐之刑曰法。杀戮无辜,爰始淫为劓、刵、椓、黥。越兹丽刑并制,罔差有辞。

"民兴胥渐,泯泯棼棼,罔中于信,以覆诅盟。虐威庶戮,方告无辜于上。上帝监民,罔有馨香德,刑发闻惟腥。

"皇帝哀矜庶戮之不辜,报虐以威,遏绝苗民,无世在下,乃命重黎,绝地天通,罔有降格。群后之逮在下,明明棐常,鳏寡无盖。

"皇帝清问下民鳏寡有辞于苗。德威惟畏,德明惟明。乃命三后,恤功于民。伯夷降典,折民惟刑;禹平水土,主名山川;稷降播种,农殖嘉谷。三后成功,惟殷于民。士制百姓于刑之中,以教祗德。

"穆穆在上,明明在下,灼于四方,罔不惟德之勤,故乃明于刑之中,率乂于民棐彝。典狱非讫于威,惟讫于富。敬忌,罔有择言在身。惟克天德,自作元命,配享在下。”

王曰:“嗟!四方司政典狱,非尔惟作天牧?今尔何监?非时伯夷播刑之迪?其今尔何惩?惟时苗民匪察于狱之丽,罔择吉人,观于五刑之中;惟时庶威夺货,断制五刑,以乱无辜,上帝不蠲,降咎于苗,苗民无辞于罚,乃绝厥世。”

王曰:“呜呼!念之哉!伯父、伯兄、仲叔、季弟、幼子、童孙,皆听朕言,庶有格命。今尔罔不由慰日勤,尔罔或戒不勤。天齐于民,俾我一日,非终惟终,在人。尔尚敬逆天命,以奉我一人!

虽畏勿畏,虽休勿休,惟敬五刑,以成三德。一人有庆,兆民赖之,其宁惟永。”

The Marquis of Lü on Punishments

When the king had occupied the throne till he reached the age of a hundred years, he gave great consideration to the appointment of punishments, in order to deal with (the people of) the four quarters.

The king said, “According to the teachings of ancient times, Chi You was the first to produce disorder, which spread among the quiet, orderly people, till all became robbers and murderers, owl-like and yet self-complacent in their conduct, traitors and villains, snatching and filching, dissemblers and oppressors.

Among the people of Miao, they did not use the power of goodness, but the restraint of punishments. They made the five punishments engines of oppression, calling them the laws. They slaughtered the innocent, and were the first also to go to excess in cutting off the nose, cutting off the ears, castration, and branding. All who became liable to those punishments were dealt with without distinction, no difference being made in favour of those who could offer some excuse. The people were gradually affected by this state of things, and became dark and disorderly. Their hearts were no more set on good faith, but they violated their oaths and covenants. The multitudes who suffered from the oppressive terrors, and were (in danger of) being murdered, declared their innocence to Heaven. God surveyed the people, and there was no fragrance of virtue arising from them, but the rank odour of their (cruel) punishments.

The great Emperor compassionated the innocent multitudes that were (in danger of) being murdered, and made the oppressors feel the terrors of his majesty. He restrained and (finally) extinguished the people of Miao, so that they should not continue to future generations. Then he commissioned Zhong and Li to make an end of the communications between earth and heaven; and the descents (of spirits) ceased. From the princes down to the inferior officers, all helped with clear intelligence (the spread of) the regular principles of duty, and the solitary and widows were no longer overlooked. The great Emperor with an unprejudiced mind carried his enquirers low down among the people, and the solitary and widows laid before him their complaints against the Miao. He awed the people by the majesty of his virtue, and enlightened them by its brightness. He thereupon charged the three princely (ministers) to labour with compassionate anxiety in the people's behalf. Bo-yi delivered his statutes to prevent the people from rendering themselves obnoxious to punishment; Yu reduced to order the water and the land, and presided over the naming of the hills and rivers; Ji spread abroad a knowledge of agriculture, and (the people) extensively cultivated the admirable grains. When the three princes had accomplished their work, it was abundantly well with the people. The Minister of Crime exercised among them the restraint of punishment in exact adaptation to each offence, and taught them to reverence virtue. The greatest gravity and harmony in the sovereign, and the greatest intelligence in those below him, thus shining forth to all quarters (of the land), all were rendered diligent in cultivating their virtue. Hence, (if anything more were wanted), the clear adjudication of punishments effected the regulation of the people, and helped them to observe the regular duties of life. The officers who presided over criminal cases executed the law (fearlessly) against the powerful, and (faithfully) against the wealthy. They were reverent and cautious. They had no occasion to make choice of words to vindicate their conduct. The virtue of Heaven was attained to by them; from them was the determination of so great a matter as the lives (of men). In their low sphere they yet corresponded (to Heaven) and enjoyed (its favour)."

The king said, "Ah! you who direct the government and preside over criminal cases through all the land, are you not constituted the shepherds of Heaven? To whom ought you now to look as your pattern? Is it not to Bo-yi, spreading among the people his lessons to avert punishments? And from whom ought you now to take warning? Is it not from the people of Miao, who would not examine into the circumstances of criminal cases, and did not make choice of good officers that should see to the right apportioning of the five punishments, but chose the violent and bribe-snatchers, who determined and administered them, so as to oppress the innocent, until God would no longer hold them guiltless, and sent down calamity on Miao, when the people had no plea to allege in mitigation of their punishment, and their name was cut off from the world?"

The king said, "Oh! lay it to heart. My uncles, and all ye, my brethren and cousins, my sons and my grandsons, listen all of you to my words, in which, it may be, you will receive a most important charge. You will only tread the path of satisfaction by being daily diligent; —do not have occasion to beware of the want of diligence. Heaven, in its wish to regulate the people, allows us for a day to make use of punishments. Whether crimes have been pre-meditated, or are unpremeditated , depends on the parties concerned; —do you (deal with them so as to) accord with the mind of Heaven, and thus serve me, the One man.

Though I would put them to death, do not you therefore put them to death; though I would spare them, do not you therefore spare them. Reverently apportion the five punishments, so as fully to exhibit the three virtues. Then shall I , the One man , enjoy felicity; the people will look to you as their sure dependance; the repose of such a state will be perpetual."

(理雅各 英译)[①]

【赏析三】

《山海经》节选

后稷是播百谷。稷之孙曰叔均,是始作牛耕。大比赤阴,是始为国。禹、鲧是始布土,均定九州。炎帝之妻、赤水之子听訞生炎居,炎居生节并,节并生戏器,戏器生祝融,祝融降处于江水,生共工,共工生术器,术器首方颠,是复土穰,以处江水。共工生后土,后土生噎鸣,噎鸣生岁十有二。洪水滔天,鲧窃帝之息壤以堙洪水,不待帝命。帝令祝融杀鲧于羽郊。鲧复生禹,帝乃命禹卒布土,以定九州。

(《山海经》第十八卷)

【今译】

后稷播种百谷。稷的孙子叫叔均,发明了用牛耕作。大比赤阴开始建

① Shang Shu 尚书(汉英对照)[M].周秉钧,译.长沙:湖南人民出版社,2013:5.

立国家。禹、鲧开始划分疆土,定为九州。炎帝的妻子,赤水子听沃生炎居,炎居生节并节并生戏器,戏器生祝融,祝融下住在江水,生共工,共工生术器。术器的头顶是平的,他收复了被占的土地,呆在江水。共工生后土后土生噎鸣,噎鸣生了十二个以岁命名的儿子。洪水滔天,鲧偷窃了天帝的息壤用来堵塞洪水,事先没有得到天帝的许可。天帝令祝融把鲧杀死在羽郊。鲧生下禹,天帝便命禹最终划定骚土,定为九州。

(陈成 译注)①

Houji sowed a hundred grains. The grandsons of Houji was Shujun, who was the first one to plough farm fields with oxen. Dabichiyin was the first to rule a kingdom.King Dayu and Gun were the first to delimit a boundary and settled the territory for nine provinces of the country. Tingyao, daughter of Chishui, was the wife of Emperor Yandi. She gave birth to Yanju. Yanju gave birth to Jiebing. Jiebing gave birth to Xiqi. Xiqi gave birth to Zhurong who came down on earth and lived by the Yangtze River. Zhurong gave birth to Gonggong. Gonggong gave birth to Shuqi whose head was flat. Shuqi returned to where Zhurong lived and stayed by the Yangtze River. Gonggong gave birth to Houtu. Houtu gave birth to Yeming. Yeming gave birth to twelve sons who were named after twelve months. The great flood inundated the whole continent. Gun stole from the God of Heaven of the divine soil to curb the flood. As Gun refused to wait for the God of Heaven to allow him to do so, the God of Heaven ordered Zhurong to kill him near Mount Yushan. Later Gun gave birth to King Dayu. The God of Heaven then ordered King Dayu to delimit a boundary and settled the territory for nine provinces of the country.

(《山海经·海内经》,王宏、赵峥 译)

① The Classic of Mountains and Seas 山海经[M]. 陈成,王宏,赵峥,译. 长沙:湖南人民出版社,2010.

[赏析四]

《淮南子》节选①

“天子衣黑衣，乘铁骊。服玄玉，建玄旗，食麦与彘，服八风水，爨松燧火。北宫御女黑色，衣黑采，击磬石。其兵铩，其畜彘。朝于玄堂右个。命有司大傩，旁磔，出土牛。命渔师始渔，天子亲往射渔，先荐寝庙。令民出五种，令农计耦耕事，修耒耜，具田器。命乐师大合吹而罢。乃命四监收秩薪，以供寝庙及百祀之薪燎。”

（《淮南子·时则训》）

The Son of Heaven is in a black robe, and black horses are used to pull the carriage. He also wears a black jade, and his carriage is decorated with a black banner. He eats millet along with pork and drinks dew blown by winds from all the eight directions. Dried pine tree branches are used for cooking and flint stone is used to ignite the cooking fire. The maids of the Northern Palace dress in black and wear black silk shawls to strike the Pan (referring to a kind of musical instrument made of stone in ancient China). The representative weapon of this month is a long spear, and the representative animal is a pig. The Son of Heaven holds court at the easterly end of the north-facing hall named “Xuan Tang”. The Son of Heaven orders officials in charge of ceremonies to arrange a magnificent sacrifice to exorcise plague-causing demons, and livestock offerings are dissected alive to drive away ominous ghosts. The Son of Heaven orders the official in charge of fishery to start catching fish, and he will go in person to spear fish. Then he will taste the newly caught fish after he has first offered them at the

---

① 刘安. Huai Nan Zi Vol. Ⅰ, Ⅱ, Ⅲ. 淮南子(全三册)[M]. 牟爱鹏，翟江月，译. 桂林：广西师范大学出版社，2010.

ancestor temple. The people are ordered to store some ice in cellars. After that is done, officials are ordered to notify farmers to prepare crop seeds. Officials in charge of agriculture are ordered to prepare for the spring ploughing, repair ploughs and plough shares and purchase other farming tools if necessary. The chief court musician is ordered to perform a magnificent concert to celebrate the end of the year's training.

(牟爱鹏、翟江月 译)

昔舜耕于历山,期年而田者争处墝埆,以封壤肥饶相让;钓于河滨,期年而渔者争处湍濑,以曲隈深潭相予。当此之时,口不设言,手不指麾,执玄德于心,而化驰若神。使舜无其志,虽口辩而户说之,不能化一人。是故不道之道,莽乎大哉!

(《淮南子·原道训》)

【今译】

从前舜在历山耕种,一年后种田的人争相耕种贫瘠的土地,把土壤肥沃的地方让给别人;他在黄河岸边钓鱼,一年后打鱼的人争相在水流湍急的地方捕鱼,把水湾和深潭让给别人。正当这时,嘴里不用说什么话,手无须指挥,内心怀藏着玄德,教化就能传播开来达到神奇的效果。假使舜没有这样的志向,即便他挨家挨户辩论游说,也不能感化一个人。因此不能用言语表达的道,浩大无边啊!

(翟江月 译)

Previously, Shun farmed in Mount Li, a year later, farmers vied with one another for sterile fields and left fertile land to others, Shun angled along the Yellow River, a year later, fishermen vied with one another to go fishing at torrid places, and left placid water areas and ponds to others. At that time, Shun did not need to say anything verbally or make any gesture with his hands to direct the people. He just bore the Profound and Dark Virtue in the heart, the moral education he advocated grew and obtained supernatural results. Suppose Shun did not bear such ideals, although he had advised people from door to door, he could not influence even one person. Hence, the Tao that cannot be articulated with words is vast and boundless.

（牟爱鹏、翟江月 译）

练　习

1.段落翻译。

1)《四时纂要》:"茶子于寒露候收晒干,以湿沙土拌匀,盛筐笼内,穰草盖之。不尔,即冻不生。至二月中取出,用糠与焦土种之。于树下或背阴之地开坎,圆三尺,深一尺,熟劚,著粪和土,每坑下子六七十颗,覆土厚一寸许,相离二尺,种一丛。性恶湿,又畏日,大概宜山中斜坡、峻坂走水处;若平地,须深开沟垄以泄水。三年后,方可收茶。"(《续茶经》陆廷灿)

2)夫能理三苗,朝羽民,徙裸国,纳肃慎,未发号施令而移风易俗者,其唯心行者乎?法度刑罚,何足以致之也!是故圣人内修其本,而不外饰其末,保其精神,偃其智故。(《淮南子·原道训》)

3)关于农业生产,《海内经》载:"后稷是播百谷","叔均,是始作牛耕"。《大荒北经》载:"叔均乃为田祖。"关于手工业,《海内经》载:"义均是始为巧锤,是始作下民百巧。"(《山海经》前言)

2.辨析以下你认为最优的译文,讨论农业典籍的翻译技巧。

凡种荻蔗,冬初霜将至,将蔗砍伐,去杪与根,埋藏土内(土忌洼聚水湿处)。雨水前五六日,天色晴明即开出、去外壳,砍断约五六寸长,以两节为率。<u>密布地上,微以土掩之,头尾相枕,若鱼鳞然。两芽平放,不得一上一下,致芽向土难发。</u>芽长一二寸,频以清粪水浇之。俟长六七寸,锄起分栽。

(《天工开物·甘嗜》)

译文一: The best time to plant sugar canes is the early winter before the coming of frost. The sugar canes are uprooted and buried by earth (avoid low ground or places where water gathers) after removing the tops and roots. Five or six days before the solar term of Rain Water in the following spring, the sugar canes are dug out of the earth on sunny days. After the bark is peeled off, the canes are cut into sections five or six *cun* long leaving two joints to a section. Then all the sections will be laid on the ground next to each other and then covered with earth. The ends of the sections overlap like fish scales. The two buds on each section of the sugar cane should be put in a level position since it is difficult for them to sprout when the buds are covered by each other. When the buds grow to one or two *cun* high, they should be watered with liquid manure frequently. When the young plants are six or seven *cun* high, it is time to transplant them.

(王义静、王海燕、刘迎春 译)

译文二: When planting sugarcane, as winter approaches and frost is imminent, the canes should be harvested, with the tops and roots removed, and buried in the soil (avoiding damp and water-collecting lowlands). About five or six days before the spring rain, when the weather is clear, they should be uncovered, the outer shell removed, and cut into sections about five or six inches long, with two nodes as a standard length. They should be densely arranged on the ground, lightly covered with soil, with the ends touching like fish scales. The buds should be placed flat, not one on top of the other, to prevent the buds from being buried in the soil and having difficulty sprouting. When the buds grow one or two inches, they should be frequently watered with clear manure water. Once they have grown six or seven inches, they should be hoed up and transplanted.

第二部分

农业科技典籍翻译的方法

第六章　农业科技典籍词汇的翻译

◎本章学习目标◎

1. 了解农业科技典籍词汇的特点
2. 理解术语、文化负载词、数量词的含义
3. 掌握农业科技典籍词汇的翻译方法

英国语言学家威尔金斯(D. A. Wilkins)说过:“没有语法,表达甚少;没有词汇,表达为零。”词汇是语义构成的基本单位,也是翻译的基础。农业科技典籍中的词汇不仅包含着大量的农业科技信息,还承载了厚重的中国传统文化,与中华民族的农耕文明和哲理思想紧密相连。

典籍多以文言文书写,词汇常由单一汉字构成,言简意赅,信息密集,相较于普通科技词汇来说,在理解和翻译上的难度更大。翻译时需对农典进行通俗化今释,也就是语内翻译,必要时需采用脚注、评注、插图、扩充性介绍等深度翻译的方法。举例如下:

煮胶法:煮胶要用二月、三月、九月、十月,余月则不成。热则不凝无饼;寒则冻瘃,令胶不黏。

Glue-making ought to be done in the 2nd, 3rd, 9th and 10th months (about March, April, October and November); success could not be assured in other months. In too hot seasons, congealment will be difficult, and no cake will be formed; in too cold season, glue will have chilblains* and does not stick.

* "chilblains" of glue means cracking and exudating while still soft and yielding.

(《齐民要术》卷九《煮胶第九十》,石声汉 译)

原文记述了煮胶的最佳时间及原因,煮胶要在二月、三月、九月、十月;其余月份不行。天热不凝固,没有胶饼;天冷,冻了离浆开裂,胶没有黏力。术语"瘃"(zhú)是人体冻伤后,充血肿大、皮肤裂开淌水的情形。胶冻在低温中,发生"离浆"开裂的现象,有些和冻瘃相似,但并不完全相同。石声汉将其翻译为chilblains,并加注进行了解释。此外,这一例子还涉及到农典中月份的理解和翻译问题,将在本章第三节详细讲解。

第一节　术语的翻译

翻译任何文本,首先要确立正确的术语(terminology)。正如严复所言:"今夫名词者,译事之权舆也,而亦为之归宿。"术语翻译不正确,译文难以忠实传达原文的意思。尤其是科技文本,因其规范性和严密性的特点,对术语翻译质量的要求更高。

农业科技典籍中使用了大量的术语,如农作物名称、农业生产和加工过程、生产工具、节气等,这些术语既具有科技术语的一般属性,用于准确传达具体的农业科学和技术知识,同时也具有中华思想文化术语的独特性。

【例1】

得天时,则不务而自生;得人心,则不趣而自劝。

At the right time, the crops will flourish without great effort. With the support of the people, the emperor can rule the country without undue force, as his people will be willing to cultivate themselves.①

（《韩非子·功名》）

【例2】

顺天时，量地利，则用力少而成功多。任情反道，劳而无获。入泉伐木，登山求鱼，手必虚；迎风散水，逆坂走丸，其势难。

Follow the appropriateness of the season, consider well the nature and conditions of the soil, then and only then least labour will bring best success. Rely on one's own idea and not on the orders of nature, then every effort will be futile. To enter a pool looking for trees to fell or to seek fish on mountains, one is bound to come back empty-handed. To spray water against a head wind or to try to roll a ball up a slope, the circumstances spell difficulty.

（《齐民要术》卷一《种谷第三》，石声汉 译）

以上两例中的术语“天时”，一方面指农业生产必须遵循的自然规律，另一方面也体现了中国古代思想中“顺自然，适其时事”的天时观。

术语翻译的标准在于“准确性，可读性，透明性”三者。术语翻译要准确表达原文的意义，不能误导读者，同时还必须具有可读性，便于使用，准确而不可读就不成其为翻译。透明性是指读者能从译名轻松辨认出源词，能轻松回译。准确性是第一位的，可读性和透明性不能以牺牲准确性为代价。②

① 译文来源：中国特色话语对外翻译标准化术语库“天时”词条。

② 姜望琪．论术语翻译的标准[J].上海翻译，2005(S1)：80-84.

一、直译

【例3】

春气未通,则土历适不保泽,终岁不宜稼,非粪不解。

In the spring, when the breath of the earth has not come through, the soil will be lumpy when ploughed; it will be unable to retain moisture, and thus does not support the growth of crop plants for the whole year to come, unless heavily manured.

(《氾胜之书》,石声汉 译)

【例4】

冬雨雪,止。辄以蔺之,掩地雪,勿使从风飞去。后雪,复蔺之;则立春保泽,冻虫死。来年宜稼。

Upon every pause of snowfall, roll down so as to catch any snow on the ground surface and stop its drifting away by wind. Roll down the later snowfalls in the same way. Moisture of the soil is thus secured for the spring to come, insects will be killed by the freezing of the soil water and good crops for the harvest will thus be warranted.

(《氾胜之书》,石声汉 译)

直译即严格按照原文的意思进行翻译。在以上两例中,术语"保泽"的意思是使土壤保持一定的水分,直译为 to retain / secure moisture of the soil,明确而直接地传达了该术语的含义。类似的例子还有:

【例5】

三月榆荚时,雨,高地强土,可种禾。

In the 3rd month, when elm-trees are fruiting, spiked millets may be sown in heavy soils on high land whenever it rains.

(《氾胜之书》,石声汉 译)

【例6】

到榆荚时,注雨止,候土白背,复锄。

During the fruiting of elm-trees, with every downpour of rain, watch the ground. When the surface turns pale, hoe again.

(《氾胜之书》,石声汉 译)

"榆荚"是榆树结的果,"榆荚时"指榆树结果的时候,用作古代农耕的一种时间标志,这里均采取了直译。

【例7】

"深耕"二字不可施之菽类,此先农之所未发者。

Deep ploughing is not suitable for beans, but it was not realized by farmers before now.

(《天工开物·乃粒第一卷》,王义静、王海燕、刘迎春 译)

二、意译

【例8】

粟、黍、穄、粱、秫,常岁岁别收:选好穗纯色者,劁才雕反刈,高悬之。至春,治取别种,以拟明年种子。

Seed corns for spiked millet, ordinary and glutinous panicled millets, ordinary and glutinous Setaria, are always to be collected separately every year. Pick out plump ears uniform in colour; cut them down and hang up to dry. Next spring, thresh those choice ears to sow in a separate parcel of land for reproducing in the year to come.

(《齐民要术》卷一《收种第二》,石声汉 译)

无论粟、黍子、穄子、粱米、秫米,总要年年分别收种:选出长得好的穗子,颜色纯洁的,割下来,高高挂起。到第二年春天打下来,另外种下,预备明年作种用。原文中的"粟、黍、穄、粱、秫"是地道的本土农作物,在英语中没有对等物,翻译时依照农作物的主要特点进行了意译。类似的例子还有:

【例9】

穄青喉,黍折头。

For common panicled millet, reap while the neck is still green; for glutinous panicled millet, do not cut until the ears are lean.

(《齐民要术》卷二《黍穄第四》,石声汉 译)

【例10】

粱秫收刈欲晚,早刈损实。

Common or glutinous Setaria should be reaped late. Early reaping means unfilled corn.

(《齐民要术》卷二《粱秫第五》,石声汉 译)

【例11】

豆种亦有二:一曰摘绿,荚先老者先摘,人逐日而取之;一曰拔绿,则至期老足,竟亩拔取也。

There are two kinds of mung beans. One is the ripe-and-pick type, which means farmers pick the ripened pods on a daily basis, while the other type is the ripe-and-pull type, which means farmers harvest the bean stalks when all the beans are ripe.

(《天工开物·乃粒第一卷》,王义静、王海燕、刘迎春 译)

三、音译

在石声汉自译《齐民要术概论》中,还采用了"汉字+音译+注"的方法,翻译介绍了以下几种农作物术语：

黍 Shu (glutinous panicled millet)

穄 Chi (non-glutinous panicled millet)

粱 Liang (ordinary Setaria)

秫 Shuh (glutinous Setaria)

麦 Mai (wheat)

稻 Tao (rice)

秫稻 Shuh-tao (glutinous rice)

需要注意的是，在石声汉英译农典中，均使用的是威妥玛式拼音法（Wade-Giles romanization），简称威氏拼音法。常见的诸如功夫（Kungfu）、清明（Chingming）、太极（Taichi）、宫保鸡丁（Kungpao Chicken）等词的翻译，使用的便是威氏拼音法。

我国《汉语拼音方案》于1958年发布，并于1982年成为国际标准ISO 7098。汉语拼音是现今国际普遍承认的现代标准汉语拉丁转写标准，“但用威妥玛音译的中国专名大量存在于西方文献中，即便在现今的西方学界也颇具影响。所以翻译中国典籍中的专名术语时，有通用译名，则遵照已有译名；无通用译名，则采用汉语拼音音译+括号注译（威妥玛音译+解释性翻译），以减少西方读者的理解障碍。”①

【例12】

<u>胡麻</u>，直是今油麻，更无他说。予已于《灵苑方》论之，其角有六棱者、有八棱者；中国之麻今谓之<u>大麻</u>是也。有实为<u>苴麻</u>，无实为<u>枲麻</u>，又曰牡麻。张骞始自大宛得油麻之种，亦谓之<u>麻</u>，故以“<u>胡麻</u>”别之，谓汉麻为“<u>大麻</u>”也。

① 张保国，周鹤. 石声汉的农学典籍译介模式及其启示[J]. 解放军外国语学院学报，2022，45(5)：123.

Huma is today's sesame and there is no other name for it. As I have described in my *Lingyuan Prescriptions*, its pod has six or eight ridges. Dama is actually hemp grown in Central Plains. Those that can bear fruit are called "ju" while those that cannot bear fruit are called "xi" or male hemp. The sesame first taken back by Zhang Qian from Dawan, formerly a remote state in the west, is also called "ma". Later sesame is called "huma" while hemp is called "dama".

(《梦溪笔谈·药议卷二十六》,王宏、赵峥 译)

【例13】

凡麦有数种。小麦曰来,麦之长也。大麦曰牟、曰穬。

Wheat, in the broad sense, is of different types, but in the narrow sense, it is called lai, which is the main type of wheat. Barley is called mou or kuang.

(《天工开物·乃粒第一卷》,王义静、王海燕、刘迎春 译)

四、加注

有些术语无法单纯用直译、意译或音译达意,则需要附加解释性的注释,以消除跨文化理解上的障碍。注释可以是文外注或文内注,也可以两者结合使用,还可以结合音译使用。比如:

1. 文外注

【例14】

春,冻解,地气始通,土一和解。夏至,天气始暑,阴气始盛,土复解。夏至后九十日,昼夜分,天地气和。——以此时耕田,一而当五,名曰"膏泽",皆得时功。

In spring time, after thawing, the breath of the earth* comes through, so the soil breaks up for the first time. With the summer solstice, the weather begins to become hot, and the yin breath* strengthens, the soil breaks up again. Ninety days after summer solstice, duration of the day equals to that of the night, and the breath of the heaven* harmonises with that of the earth. To plough in these proper seasons, one operation is worth five. Such conditions are denoted as kao ts'eh**, therein lies the benefit of appropriate timing.

* Breath of the earth or yin breath means the complex conditions of low temperature and high humidity of the soil and the reverses of the air. Breath of the heaven, on the other hand, indicates warm and dry conditions prevailing under sunshine.

** Kao ts'eh 膏泽, literally equivalent to "greasy moisture".

(《氾胜之书》,石声汉 译)

2. 音译+文内注

【例 15】

何谓八风？东北曰炎风，东方曰滔风，东南曰熏风，南方曰巨风，西南曰凄风，西方曰飂风，西北曰厉风，北方曰寒风。

What are the eight different kinds of winds? The northeasterly is called Yan Feng (which means very hot wind). The easterly is called Tao Feng (which means strong wind originating from the sea). The southeasterly is called Xun Feng (which means gentle wind). The southerly is called Ju Feng (which means very heavy wind). The southwesterly is called Qi Feng (which means very sharp and unmerciful wind). The westerly is called Liu Feng (which means harmfully heavy and sharp wind). The northwesterly is called Li Feng (which means bitterly sharp wind). And the northerly is called Han Feng (which means fiercely chilly wind).

(《吕氏春秋》,翟江月 译)

【例16】

秋无雨而耕,绝土气,土坚垎,名曰“腊田”。及盛冬耕,泄阴气,土枯燥,名曰“脯田”。脯田与腊田,皆伤。

If one ploughs in autumn when it does not rain, the breath of the earth is cut off, and the soil will be hard and cloddy. This is called “lah-t'ien” (bacony field). If one ploughs in a severe winter, the yin breath of the earth is broached and the soil will be dry and parched. This is called “fu-t'ien” (jerked field). Both bacony and jerked fields are harmed ground.

(《氾胜之书》,石声汉 译)

【例17】

王祯农桑通诀曰:木棉谷雨前后种之。立秋时,随获随收。其花黄如葵。其根独而直。其树不贵乎高长,其枝干贵乎繁衍。不由宿根而出,以子撒种而生。所种之子,初收者未实,近霜者又不可用,惟中间时月收者为上。须经日晒燥,带棉收贮。临种时再晒,旋碾即下。

Wang Ching (王祯) says that cotton seeds should be sown about the commencement of the kuh-yü term (April 20th); and gathered as the cotton ripens in the lih-tsiú (Aug 8th). The flower is yellow like the Althea, its single root is straight; its excellence does not consist in its height and expanse, but in the branches and leaves being bushy and numerous. It does not sprout out from the last year's roots, but the seeds must be annually sown; the seeds first gathered are not fully ripe, and those collected near the hoar-frosts are useless; the best are collected in the intervening season, and should be dried in the sun, and laid up with the cotton around them, drying them again when about to sow them, and then separating the kernels in the gin.

(《农政全书》卷三十五《蚕桑广类·木棉》,C. Shaw 译)

3. 音译+汉字+文内注

Table 5: *Contents of the* Nung Cheng Chhüan Shu

Chapters 1-3:	*Nung Pen* 農本 (Fundamentals of Agriculture): quotations from the classics, etc. on importance of encouraging agriculture
4-5:	*Thien Chih* 田制 (Field Systems): land distribution, field management
6-11:	*Nung Shih* 農事 (Agricultural Tasks): clearing land, tilling, etc., including a detailed exposition of settlement schemes
12-20:	*Shui Li* 水利 (Water Control): various methods of irrigation and types of irrigation equipment, including 19-20: *Thai Hsi Shui Fa* 泰西水法 on Western irrigation equipment
21-24:	*Nung Chhi Thu Phu* 農器圖譜 (Illustrated Treatise on Agricultural Implements): largely based on the same section in *WCNS*
25-30:	*Shu I* 樹藝 (Horticulture): vegetables and fruit
31-34:	*Tshan Sang* 蠶桑 (Sericulture)
35-36:	*Tshan Sang Kuang Lei* 蠶桑廣類 (Further Textile Crops): cotton, hemp
37-40:	*Chung Chih* 種植 (Silviculture)
41:	*Mu Yang* 牧養 (Animal Husbandry)
42:	*Chih Tsao* 製造 (Culinary Preparations)
43-60:	*Huang Cheng* 荒政 (Famine Control): 43-45: administrative measures; 46-60: *Chiu Huang Pen Tshao* 救荒本草 (Famine Flora)

李约瑟(Joseph Needham)主编《中国科学技术史》中的《农政全书》目录

【例 18】

凡田欲早晚相杂。防岁道有所宜。有闰之岁，节气近后，宜晚田；然大率欲早。

In planning the cultivation, take an assortment of lands for one crop to provide for possibly different courses of climate in the year. For those years with an intercalary month, field operations might be begun a little late, for the 节气 tsie ch'i, i.e. the twenty four sub-seasons of a year after that month usually lag a few days. But earliness is generally recommended.

（《齐民要术》卷一《种谷第三》，石声汉译）

在 2022 年北京冬奥会的官方翻译中，“二十四节气”被译为 The 24 solar terms，每个节气的对应译名如下：

立春	Beginning of Spring	立秋	Beginning of Autumn
雨水	Rain Water	处暑	End of Heat

续表

惊蛰	Awakening of Insects	白露	White Dew
春分	Spring Equinox	秋分	Autumn Equinox
清明	Pure Brightness	寒露	Cold Dew
谷雨	Grain Rain	霜降	Frost's Descent
立夏	Beginning of Summer	立冬	Beginning of Winter
小满	Grain Buds	小雪	Minor Snow
芒种	Grain in Ear	大雪	Major Snow
夏至	Summer Solstice	冬至	Winter Solstice
小暑	Minor Heat	小寒	Minor Cold
大暑	Major Heat	大寒	Major Cold

4.音译+汉字+文内注+文外注

【例19】

上农夫:区,方深各六寸,间相去九寸。

For *shang nung fu** (上农夫, or lots of best land for a farm family), the ou or shallow pit should be 6 ts'un across, 6 ts'un deep, and 9 ts'un apart.

*Word by word, the term "shang nung fu" (上农夫) should be rendered as "best farmer". By reference to commentary statements given in *Hou Han Shu* (Chronicle of Later Han), I now interpret "fu" as the quota of land-lot for a farm family, as were several times used in "orders for Land-Allotment" from Han Dynasty onwards, and shang nung fu will be the best land lot received by a farm family, i.e., a parcel of field which generally needs no fallowing.

(《氾胜之书》,石声汉 译)

五、插图

在译介农耕过程、工具等术语时，借助原著插图的可视化表达，往往能更直观准确地传达农典的文本意义。以下两例均为稻田耕作工序，农耕工具将在下一节详细讲解。

【例20】

凡一耕之后，勤者再耕、三耕，然后施耙，则土耕地质匀碎，而其中膏脉释化也。

Industrious farmers plough over the fields two or three times before they use the harrow to break the soil evenly and finely. In this way the fertilizers will be well distributed in the soil.

（《天工开物·乃粒第一卷》，王义静、王海燕、刘迎春 译）

耕

Loosening the soil by ploughing

耙

Breaking the soil into fine particles by harrowing

【例21】

青叶既长,则耔可施焉。……非足力所可除者,则耘以继之。

When new leaves start to grow, it is time to heap mud around the roots of the young plant (nurturing the root, which is commonly known as "foot-patting the plants")... (weeds that) cannot be broken this way and they have to be uprooted by hand.

(《天工开物·乃粒第一卷》,王义静、王海燕、刘迎春 译)

耔

Foot-patting the plant

耘

Hand weeding

练　习

1.翻译以下典籍原文,特别注意术语的翻译方法。

1)马质,禁原蚕者。(《周礼》)

2)仲冬斩阳木,仲夏斩阴木。(《周礼》)

3)春,地气通,可耕坚硬强地黑垆土。辄平摩其块,以生草;草生,复耕之;天有小雨,复耕。和之,勿令有块,以待时。所谓"强土而弱之"也。(《氾胜之书》)

4)杏始华荣,辄耕轻土弱土。望杏花落,复耕;耕辄劳之。草生,有雨,泽,耕重劳之。土甚轻者,以牛羊践之。如此,则土强。此谓"弱土而强之"也。(《氾胜之书》)

5)中农夫:区,方七寸,深六寸,相去二尺。(《氾胜之书》)

6)下农夫:区,方九寸,深六寸,相去三尺。(《氾胜之书》)

7)虜小麦,其实大麦形,有缝。稊麦,似大麦,出凉州。旋麦,三月种,八月熟,出西方。赤小麦,赤而肥,出郑县。山提小麦,至黏弱,以贡御。有半夏小麦,有秃芒大麦,有黑穬麦。(《广志》)

8)凡稻田刈获不再种者,土宜本秋耕垦,使宿稿化烂,敌粪力一倍。或秋旱无水及怠农春耕,则收获损薄也。凡粪田若撒枯浇泽,恐淋雨至,过水来,肥质随漂而去。谨视天时,在老农心计也。(《天工开物》)

9)便民图纂曰:棉花,谷雨前后,先将种子,用水浸片时,漉出,以灰拌匀。候芽生,于粪地上每一尺作一穴,种五七粒。待苗出时,密者芟去,止留旺者二三科。频锄,时常掐去苗尖,勿令长太高。若高,则不结子,至八月间收花。(《天工开物》)

2.将以下两段话译成英文。

1)二十四节气起源于黄河流域,远在春秋时期,中国古代先贤就定出仲春、仲夏、仲秋和仲冬等四个节气,以后不断地改进和完善,到秦汉年间,二十四节气已完全确立。二十四节气是我国劳动人民独创的文化遗产,它能反映季节的变化,指导农事活动,影响着千家万户的衣食住行。二十四节气为中国大众所普遍接受,日常生活中随处可见二十四节气的影响,一些节气和民间文化相结合,已经成为人们的固定节日。最著名的清明、立春、立夏、冬至都融入了节日的氛围。

2)《齐民要术》是北魏贾思勰所撰的一部农学书籍,是现存最完整的古农书。"齐民"指平民百姓,"要术"指谋生方法。约成书于北魏末年(533—544),全书十卷,十一万字,分别论述各种农作物、蔬菜、果树、竹木的栽培,家畜、家禽的饲养,农产品加工和副业等,系统地总结了六世纪以前我国黄河中下游地区农牧业生产经验、食品的加工与贮藏、野生植物的利用等,反映了以农为本、多种经营、重视购销的思想,对古代农学的发展产生了重要影响。

3. 下文记述了稻谷加工过程中击禾、轧禾和风车三道工序。根据释文进行翻译,并翻译插图中的术语。

凡稻刈获之后,离稿取粒。束稿于手而击取者半,聚稿于场而曳牛滚石以取者半。凡束手而击者,受击之物或用木桶,或用石板。收获之时雨多霁少,田、稻交湿,不可登场者,以木桶就田击取。晴霁稻干,则用石板甚便也。

凡服牛曳石滚压场中,视人手击取者力省三倍。但作种之谷,恐磨去壳尖,减削生机,故南方多种之家,场禾多借牛力,而来年作种者,宁向石板击取也。凡稻最佳者,九穰一秕。倘风雨不时,耘籽失节,则六穰四秕者容有之。凡去秕,南方尽用风车扇去。北方稻少,用扬法,即以扬麦、黍者扬稻,盖不若风车之便也。

(《天工开物·粹精第四卷》)

湿田击稻

稻场击稻

牛碾

风车

【释文】

水稻收割之后，要脱秆取粒。手握一把稻秆击取稻粒的占一半，将稻都放在场上以牛拉石磙碾取稻粒的也占一半。以手击取稻粒，被击之物或用木桶，或用石板。收获时如雨天多晴天少，田间和水稻都湿，则不可上场，便用木桶在田间就地击取。晴天稻干，则用石板击稻更为方便。

用牛拉石磙压场脱粒，比以手击稻省力三倍。但留作种子的稻谷，恐怕会磨去稻壳壳尖而减少发芽机会，所以南方种稻多的农家在场上脱谷多借牛力，而来年作稻种的则宁取用石板击取的。最好的稻谷每十棵中有九棵是颗粒丰满的，只有一棵是谷粒不饱满的。倘风雨不调，壅根拔草不及时，则间或有六棵粒满、四棵谷粒不饱满。去掉秕子时，南方都用风车扇去。北方稻少，则用扬场的方法，就是用扬麦、扬黍的方法来扬稻，但不如风车方便。

延伸阅读

《氾胜之书》译文赏析

下文节选自石声汉自译《氾胜之书今释》，文中列举了十二种农作物的种植方法和注意事项，包括禾、黍、麦、稻、大豆、小豆、麻、枲、瓠、芋、稗、桑。

种　植

Cultivation of Certain Crop Plants

种禾无期，因地为时。

There is no fixed date for sowing spiked millet. All depends on conditions of the ground.

三月榆荚时，雨，高地强土，可种禾。

In the 3rd month, when elm-trees are fruiting, spiked millets may be sown in heavy soils on high land whenever it rains.

薄田不能粪者，以原蚕矢杂禾种种之，则禾不虫。

On poor land where manures are lacking, seeds of spiked millet may be sown in mixture with excrement of polyvoltine silkworm—bombyxine excrement protects the millet from insect pests.

稙禾，夏至后八十、九十日，常夜半候之：天有霜，若白露下，以平明时，令两人持长索，相对，各持一端，以槩禾中，去霜露。日出，乃止。如此，禾稼五谷不伤矣。

For early spiked millet, always look out sharply at midnight eighty to ninety days after the summer solstice. If there should be frost or white dew, let two persons facing each other drag a rope horizontally right through the crop to clear away the frost or dew. Stop only after sunrise. This protects crop plants from frost injuries.

黍者，暑也；种者必待暑。先夏至二十日，有雨，此时。强土可种黍；亩三升。

The glutinous millet, *shu*, means a plant of "shu" (hot weather). Therefore it should be sown only when the weather is hot. Twenty days before summer solstice is the proper time. Sow 3 *shêng* per *mu* on heavy soils when it ever rains.

黍心未生，雨灌其心，心伤无实。

Before emergence, the young spikes of glutinous millet may be injured by infiltrating raindrops and hence will set no seed.

黍心初生，畏天露。令两人对持长索，搜去其露，日出乃止。

Newly emerged spikes of glutinous millet are very sensitive to dew. Let two persons facing each other stretch a long rope through to scrape away the dew. Stop only after sunrise.

凡种黍，覆土，锄治，皆如禾法，欲疏于禾。

For glutinous millet, the thickness of soil layer to cover the seeds sown and the later hoeing and care are similar to those for spiked millet. But the dropping ought to be thinner.

凡田有六道，麦为首种。种麦得时，无不善。夏至后七十日，可种宿麦。早种，则穗强而有节；晚种，则穗小而少实。

For one field, of which there are six alternates, wheat is the first. Timely sowing of wheat insures successful harvest. Winter wheat may be sown seventy days after summer solstice. Earlier sowing gives solid ears and firm straws; late sowing gives smaller ears and less plump grains.

麦生，黄色，伤于太稠。稠者，锄而稀之。

Should young wheat plants look sickly yellow, it is the injury from overcrowding. Hoe to thin them down.

秋,锄,以棘柴耧之,以壅麦根。故谚曰:"子欲富,黄金覆。""黄金覆"者,谓秋锄麦,曳柴壅麦根也。

In autumn, hoe again, and drag bunches of thorns through the rows of wheat plots so as to bank up the roots. The proverb goes: If you wish to be rich, Bank with the golden earth yellowish. "Bank with the golden earth yellowish" means the autumn hoeing and banking up the roots by dragging thorns through the plots.

冬雨雪止,以物辄蔺麦上,掩其雪,勿令从风飞去。后雪,复如此。则麦耐旱多实。

In winter, upon every pause of snowfall, roll wheat plots to catch the snow and stop its being drifted away by wind. Repeat with later snowfalls. Wheat will be drought-tolerant and yield more grains.

至春冻解,棘柴曳之,突绝其干叶。须麦生,复锄之。到榆荚时,注雨止,候土白背,复锄。如此,则收必倍。

When it thaws in spring time, drag thorns through the wheat plots again so as to detach dry dead leaves. Wait until the wheat is green, then hoe. During the fruiting of elm-trees, with every downpour of rain, watch the ground. When the surface turns pale, hoe again. These things being done, yield will be double.

春,冻解,耕和土,种旋麦。麦生,根茂盛,莽锄如宿麦。

When it thaws in spring time, plough the soil that has become mellow and sow spring wheat. After sprouting and good growth of the root system, hoe roughly as with winter wheat.

种稻:春冻解,耕反其土。种稻区不欲大,大则水深浅不适。

To plant rice: When it thaws in spring time, plough to turn over the ground. Rice field should not be too large; in too large a rice field it is difficult to adjust the height of standing water.

三月种秔稻，四月种秫稻。

Sow ordinary rice in the 3rd month, and glutinous rice in the 4th.

冬至后，一百一十日，可种稻。稻地美，用种亩四升。

Hundred and ten days after winter solstice, rice may be sown. For good fertile land, use 4 *shêng* of seeds per *mu*.

始种，稻欲温。温者，缺其塍令水道相直。夏至后，太热，令水道错。

At earlier stages of growth, the rice plants must be kept warm. To keep warm, one should make the inlet and outlet gaps on the *shêng* directly opposite to each other. After summer solstice, it becomes intensely hot, then make the gaps distantly across.

大豆保岁易为，宜古之所以备凶年也。谨计家口数，种大豆；率：人五亩，此田之本也。

From soya beans a good crop can be easily secured even in adverse years, therefore it is quite natural for the ancient people to grow soya as a provision against famine. Calculate the acreage to be covered by soya beans for members of the whole family according to the rate of 5 *mu* per capita. This should be looked as "the basic" for farming.

三月榆荚时，有雨，高田可种大豆。土和，无块，亩五升；土不和，则益之。

In the 3rd month, when elm-trees are fruiting, sow soya beans on highland fields whenever it rains. Use 5 *shêng* of seeds per *mu* when the soil is mellow and not cloddy, but more seeds if the soil is not so.

种大豆，夏至后二十日尚可种。

As late as twenty days after summer solstice soya may still be sown.

大豆戴甲而生,不用深耕。种之上,土不可厚,才令蔽豆耳。厚则折项,不能上达,屈于土中而死。

Soya seedlings come out as though with a helmet on the top, so there is no need to plough very deep. Don't cover the seed with too much soil after sowing, only just enough to screen off. Too thick a cover renders the bean-stalk bent-necked, and seedlings may never reach above ground and die prematurely underneath.

大豆须均而稀。

Dropping of soya beans ought to be uniform and thin.

豆花,憎见日,见日则黄烂而枯焦也。

Flowers of soya bean dislike sunshine; direct exposure causes their yellowing and scorching.

小豆不保岁,难得。

From lesser beans, a good crop cannot be safely secured. It is hard to grow.

宜椹黑时注雨种,亩五升。

When mulberries are darkening, sow lesser beans with heavy downpour of rain. Drop 5 *shêng* per *mu*.

豆生布叶,锄之;生五六叶,又锄之。

Hoe when foliage leaves expand. Hoe again when 5 or 6 leaves appeared.

大豆、小豆,不可尽治也。古所以不尽治者,豆生布叶,豆有膏,尽治之,则伤膏,伤则不成。而民尽治,故其收耗折也。故曰:"豆不可尽治。"

Both soya and lesser beans should not be excessively defoliated. In old times, defoliation was rather refrained, because the people then knew that foliage leaves produce nourishment for the plants themselves. Excessive defoliation means maiming the nourishing process and induces loss in harvest. Nowadays people defoliate too much and thereby often suffer losses in cropping of seeds. Hence the saying: refrain from defoliating beans excessively.

养豆，美田亩可十石；以薄田，尚可亩收五石。

With proper care, the yield from good field may attain 10 *shih* per *mu*, from poor land, up to 5 *shih*.

种麻，预调和田。

To plant hemp: Ameliorate the ground thoroughly before sowing.

二月下旬，三月上旬，傍雨种之。麻生布叶，锄之。率：九尺一树[①]。

By the end of the 2nd and the beginning of the 3rd month, sow after a rainfall. After sprouting and expansion of the foliage leaves, hoe to thin down. Proper spacing is 2 *ch'ih* apart.

树高一尺，以蚕矢粪之，树三升。无蚕矢，以溷中熟粪粪之，亦善，树一升。

When the plants grow to 1 *ch'ih* high, manure with bombyxine excrement at the rate of 3 *shêng* per plant. Failing bombyxine excrement, use well ripened night soil from pits instead. The rate is then 1 *shêng* per plant.

天旱，以流水浇之，树五升。无流水，曝井水杀其寒气以浇之。雨泽时适，勿浇！浇不欲数。

Irrigate with flowing water, 5 *shêng* per plant at one time. Falling streaming water, sun well-water to warm it up before application. With timely rainfall and proper soil moisture, do not water. Watering should not be too frequent.

养麻如此，美田则亩五十石，及百石，薄田尚三十石。

Hemp plants thus well cared for will yield 50 *shih* up to 100 *shih* per mou with good fertile land and 30 *shih* with poor land.

---

① “九尺一树”的间距太长，与情理不合。石声汉根据《齐民要术》中的记述，在译文中改为了“二尺”。

种枲:春冻解,耕治其土。春草生布,粪田,复耕,平摩之。

To plant male hemp: Plough and drill the field in spring time when it thaws. After sprouting of weeds, apply manure and plough down, harrow to level out then sow.

种枲太早,则刚坚,厚皮多节;晚则皮不坚。宁失于早,不失于晚。

When sown too early, male hemp plants will be hard and rigid, pachydermous and knotty; too late, not hard enough. Rather too early than too late.

种瓠法:以三月,耕良田十亩,作区,方深一尺。以杵筑之,令可居泽。相去一步。区,种四实。蚕矢一斗,与土粪合。浇之,水二升;所干处,复浇之。

To plant gourd: In the 3rd month, prepare 10 *mu* of good fertile field, dig out shallow pits 1 *ch'ih* in diameter and deep, and 6 *ch'ih* apart. Pound the bottom with a club, so as to make it firm and waterproof. Place 1 dou bombyxine excrement mixed with some compost into each pit, and sow 4 seeds thereupon. Then apply 2 *shêng* water. Water again where it dries up.

著三实,以马箠殼其心,勿令蔓延;多实实细。以槁荐其下,无令亲土,多疮瘢。

After 3 fruits have been formed on one vine, knock off the tip of that vine with a long whip to stop over-growth. The more are numbers of fruits per vine, the smaller they will be in size. Place some dry straw underneath every fruit to prevent direct contact with the earth, else there will be scars on the skin.

度可作瓢,以手摩其实,从蒂至底,去其毛,不复长,且厚。八月微霜下,收取。

When it is considered good enough to make ladle with, rub with hands from bottom to stalk to depilate the fruit, thence they will grow only in thickness but no more in size. In the 8th month, after a mild frost, gather them up.

掘地深一丈，荐以槁，四边各厚一尺。以实置孔中，令底下向。瓠一行，覆上，土厚三尺。

Dig a pit 10 *ch'ih* deep, lay some dry straw on the bottom and all around the side up to 1 *ch'ih* thick. Place the fruits gathered bottom downwards in layers upon the straw, cover each layer with dry earth, 3 *ch'ih* thick.

二十日出，黄色，好，破以为瓢。其中白肤，以养猪，致肥；其瓣，以作烛，致明。

Twenty days hence, pick out the gourds. Now they should be good, yellow and ready. Half them into ladles. The white flesh within is good for feeding pigs—being highly fattening. Seeds are good for making flambeaux—burning exceedingly bright.

一本三实，一区十二实，一亩，得二千八百八十实。十亩，凡得五万七千六百瓢。瓢直十钱，并直五十七万六千文。用蚕矢二百石，牛耕工力，直二万六千文。余有五十五万。肥猪明烛，利在其外。

Each vine yields 3 fruits, so each pit yields 12 fruits, one *mu* gives 2,880 fruits, 10 *mu* produces 57,600 ladles. Each ladle is worth 10 cash, so the total price is 576,000 cash. For the bombyxine excrement used, together with cattle-service and man power, the cost is 26,000 cash. Net profit is thus 550,000 cash, exclusive of the benefit of fattening of pigs and flambeaux.

种芋法：宜择肥缓土，近水处。和柔，粪之。二月，注雨，可种芋。率：二尺下一本。

To plant taro: Fields fertile and loose, and situated near water sources should be taken. Drill and manure them. In the 2nd month, with downpours of heavy rain, plant taro 2 *ch'ih* apart.

芋生，根欲深。斸其旁，以缓其土。旱，则浇之。有草，锄之，不厌数多。治芋如此，其收常倍。

After sprouting, the roots need go in deeply, so pick around the stump to loosen the soil. Water when too dry. Hoe if weeds occur. Hoeing will never be too often. From taro-plants so diligently worked, yields will double.

稗,既堪水旱,种无不熟之时,又特滋茂盛易生。芜秽良田,亩得二三十斛。宜种之备凶年。

From water-darnel (*Panicum*), which is highly tolerant both towards drought and flood, good harvest can be secured in adverse years. It is very prolific and thus easy to grow. A yield of 20 to 30 hu per *mu* can be obtained from good field badly crowded with weeds. Therefore it is highly advisable to sow it as a provision against famines.

稗中有米,熟时,捣取米;炊食之,不减粱米。又可酿作酒。

There is a kernel in every darnel grain. The ripe grain should be pounded to obtain the kernel which, when steamed, provides a good meal not inferior to *Setaria*. It may be fermented to make a wine.

种桑法:五月,取椹著水中,即以手溃之,以水灌洗取子,阴干。

To plant mulberry trees: In the 5th month (*ca.* late June), collect ripened mulberries, soak in water, and rub them between the hands. Crush with more water to obtain cleansed seeds. Air to dry.

治肥田十亩,荒田久不耕者,尤善! 好耕治之。每亩以黍椹子各三升合种之。

Take 10 *mu* of good land—best to use long - fallowed plots—plough and clean nicely. To every *mu*, sow a mixture of 3 *shêng* of mulberry seeds with equal amount of glutinous millet.

黍桑当俱生,锄之,桑令稀疏调适。

Both mulberry and millet will germinate about the same time. Hoe the mulberry seedlings down, let them be properly apart from each other.

黍熟，获之。桑生正与黍高平，因以利镰摩地刈之，曝令燥。后有风，调，放火烧之，常逆风起火。桑至春生，一亩食三箔蚕。

When the millet ripens, reap the ears, the mulberry plants will be of the same height as the ripe millet, so just cut them both down with a sickle close to the ground. Dry them together in the open. Later, with a good breeze, set fire on the dried plants in a head wind. Next spring mulberry suckers spring out so that 1 *mu* will support 3 pan of silkworms.

第二节　文化负载词的翻译

尤金·奈达(Eugene Nida)在Linguistics and Ethnology in Translation Problems一文中提出了决定语义对等的五类文化问题：

Words are fundamentally symbols for features of the culture. Accordingly, the cultural situation in both languages must be known in translating, and the words which designate the closest equivalence must be employed. An examination of selected problems in various aspects of culture will make it possible for one to see more clearly the precise relationship of cultural information to the semantic problems encountered in descriptive linguistics. Translation problems, which are essentially problems of equivalence, may be conveniently treated under (1) ecology, (2) material culture, (3) social culture, (4) religious culture, and (5) linguistic culture.①

本节将从生态、物质、社会、宗教、语言五个方面来看农业科技典籍中文化负载词的翻译。

① Nida, Eugene. Linguistics and Ethnology in Translation Problems [J]. Word, 1945, 1(2): 194–208.

一、生态

生态文化负载词反映了中国古代的生态环境、气候特征、自然地理等,主要包括四季更迭、山川河流、花草树木等。比如:

【例1】

孟春之月:日在营室,昏参中,旦尾中。其日甲乙,其帝太皞,其神句芒。其虫鳞,其音角。律中太蔟,其数八。其味酸,其臭膻。其祀户,祭先脾。东风解冻,蛰虫始振。鱼上冰,獭祭鱼,候雁北。

In the first month of spring, the sun is at the position of Yingshi[1]. In the early evening, Can[2] appears in the middle of the southern sky; at daybreak, Wei[3] appears in the middle of the southern sky. This period of time is indicated by Jia Yi[4]. The ruler of this month is Tai Hao[5]. Its god is Gou Mang[6]; its animal, scaled creatures; its musical note, Jiao[7]; its pitch, Tai Zu[8]; its number, eight; its taste, sour; its odor, mutton-like; its sacrificial ceremony, the Window ceremony[9]; and animal spleens are to be offered first. The east wind causes ice to thaw; hibernating creatures begin to stir; fish swim to the top of the water just beneath the ice; otters begin to feed on fish; migratory wild geese fly to the north.

1 Yingshi, or Shi, is one of the twenty-eight constellations in ancient Chinese astronomy. It consists of stars in the constellation Pegasus.

2 Can is one of the twenty-eight constellations. It consists of stars in the constellation Orion.

3 Wei is another one of the twenty-eight constellations. It consists of stars in the constellation Scorpio.

4 Jia and Yi are the first two of the ten Heavenly Stems, here denoting Wood, or spring.

5 Tai Hao, also known as Fu Xi, was a legendary ruler honored as the Ruler of the East.

6 Gou Mang, the son of Tai Hao, was honored as the God of Wood Virtue.

7 Jiao, the middle note, is one of the five notes in ancient Chinese music. The five notes are Gong, Shang, Jiao, Zhi and Yu.

8 Tai Zu is one of the twelve pitches set with a bamboo pitch-pipe in ancient Chinese music.

9 The Window ceremony was one of the five kinds of sacrificial ceremonies: Window, Stove, House, Door and Road. The ancients believed that when spring comes, hibernating creatures begin to stir and come out through the Window.

（《吕氏春秋》,汤博文 译）

原文记述了早春时节万物复苏的景象,包含中国古代观测季节更迭的方法,如观察太阳的位置等。文中涉及大量古代天文、神话、声乐、祭祀等词汇,翻译时均采用了音译加文外注的方法。

【例2】

天地有始。天微以成,地塞以形。天地合和,生之大经也。以寒暑日月昼夜知之,以殊形殊能异宜说之。夫物合而成,离而生。知合知成,知离知生,则天地平矣,平也者,皆当察其情,处其形。天有九野,地有九州,土有九山,山有九塞,泽有九薮,风有八等,水有六川。

Heaven and earth have their origin. Heaven was formed by light particles; earth, by heavy and muddy particles. The conjugation of heaven and earth is the foundation of life. This can be seen in the change of hot and cold weather, the movement of the sun and the moon and in the alternation of day and night and can be explained by the difference in the shape, nature and function of things. Heaven and earth conjugate to give form to things and separate to bring them into existence. If one can understand the conjugation of heaven and earth and how things are given shape and if one can understand the separation of heaven and earth and how things are brought into existence, one will be able to know the rules governing the changes in heaven and earth. In order to know the rules governing the changes in heaven and earth, it is necessary to look into the nature and shape of things. Heaven has nine segments, and earth, nine prefectures. There are nine mountains on earth, and nine passes on the mountains. There are nine great lakes, eight types of wind and six great rivers.

(《吕氏春秋》,汤博文 译)

原文记述了天地开创的源头,以及天地山河的划分,充分体现了中国古代"和生万物"的观念。天地合和是万物生存的根本,天地自然和人类活动相互依存、相互统一。"九野""九州""九山"等词均采用了直译,《吕氏春秋》中对这些术语进行了详细解释,可参考本节末的译文赏析部分。

【例3】

古人藏书辟蠹用芸。芸,香草也,今人谓之七里香者是也。叶类豌豆,作小丛生,其叶极芬香,秋后叶间微白如粉污,辟蠹殊验,南人采置席下能去蚤虱。予判昭文馆时曾得数株于潞公家,移植秘阁后,今不复有存者。香草之类大率多异名,所谓兰荪,荪即今菖蒲是也,蕙今零陵香是也,茝今白芷是也。

Ancient people used rues to keep away moths. A rue is a kind of aromatic plant which can spread fragrant smell for seven miles. Growing in small clumps, its leaves resemble those of peas and smell exceedingly sweet. In late autumn the leaves of rues will turn a bit whitish as if they are stained with white powder. Rues are very effective in repelling moths. People in the south often put the leaves of rues under their mats to repel fleas and louses. When I worked in the

Zhaowen Library, I got several seedlings of rues from Wen Yanbo and planted them behind the Secret Stack Room. None of them was alive. Fragrant herbs often have alternative names. For instance, the so-called lansun is today's calamus while hui is today's fragrant thoroughwort. And chai, an aromatic plant mentioned in ancient books, is today's dahurian angelica.

（《梦溪笔谈·辩证卷三》,王宏、赵峥 译）

原文记述了“芸”这种香草的驱虫功能和外观特点,还介绍了几种香草植物名称的变化,荪就是现在的菖蒲,蕙就是现在的零陵香,茝就是现在的白芷。各种香草植物名称的翻译分别采取了直译、意译和音译等不同方法。

二、物质

物质文化负载词反映了中国古代的劳动生产、物质生活等,包括各种农耕工具、祭祀器皿、乐器等,如《礼记·礼运》中的“琴、瑟、管、磬、钟、鼓”,可分别译为lutes、citherns、flutes、sonorous stones、bells和drums。再比如:

【例4】

耕荒毕,以铁齿镉楱,再遍杷之。漫掷黍穄,劳亦再遍;明年,乃中为谷田。

After ploughing, level down twice the clods with an iron-toothed rake. Broadcast glutinous or ordinary panicled millets, harrow twice. The plots will be ready for spiked millet next year.

（《齐民要术》卷一《耕田第一》,石声汉 译）

原文记述了开荒后的操作:荒地耕完之后,用有尖铁齿的铁搭扒两遍。撒播一些黍子和穄子,用耢摩两遍;明年就可以用来种谷子。“镉楱”(lòu zòu)是耙的一种,即“铁搭”。王祯《农书》中认为是“人字耙”,即尖齿铁钯,主要的作用是松土。劳(lào),同"耢",是一种平整土地用的农具,由牲口拉动,可以使耕翻的土块散碎又排平。役使牲口的人,可以坐或立在耢上。

"铴榛"和"劳"是中国古代特有的农具术语,在英语中没有对等物。翻译时,前者采用了意译,后者按农具的功能译为对应的动词。同时,还可以借助原文插图,以可视化方式实现文本信息的准确传达。

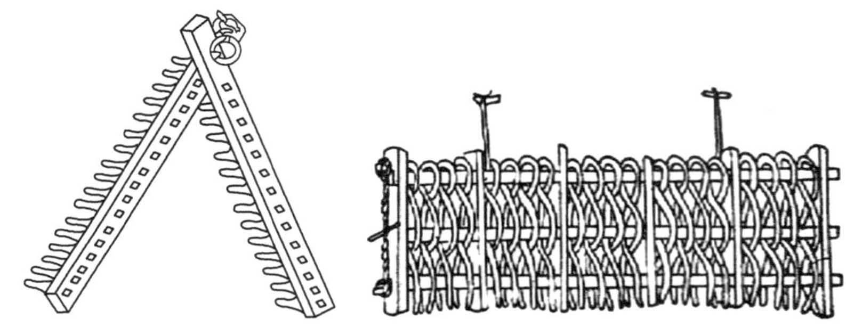

铴榛 iron-toothed rake　　劳 harrow, leveller

【例5】

士之仕也,犹农夫之耕也,农夫岂为出疆舍其耒耜哉?

Holding office for a man of service is like plowing for a farmer. When a farmer passes beyond the boundaries of a state, does he leave his plow behind?

(《孟子·滕文公下》,Irene Bloom 译)

【例6】

二之日凿冰冲冲,三之日纳于凌阴。四之日其蚤,献羔祭韭。九月肃霜,十月涤场。朋酒斯飨,曰杀羔羊。跻彼公堂,称彼兕觥,万寿无疆!

In the twelfth moon we hew out ice;

In the first moon we store it deep.

In the second we offer early sacrifice

Of garlic, lamb and sheep.

In ninth moon frosty is the weather;
In tenth we sweep and clear the threshing-floor.
We drink two bottles of wine together
And kill a lamb before the door.
Then we go up to the hall where
We raise our buffalo-horn cup
And wish our lord to live fore'er.

(《诗经·国风·豳风·七月》,许渊冲 译)

十二月凿冰声冲冲忙,正月里把冰往冰室藏。二月里取冰祭祀早,献上韭菜和羔羊。九月里降下霜,十月里清扫打谷场。两壶酒可以上飨,再杀了羔羊,登那公爷堂,举起那兕角觥,说万寿无疆!"兕"(sì)指古代犀牛一类的兽名,外形似牛,身有黑毛,头长独角。"兕觥"是用兕角做成的酒器,属于中国古代特有的物质文化,译者采用了解释性翻译,将其译为buffalo-horn cup,帮助英语读者有效理解。

【例7】

宗庙之祭,贵者献以爵。贱者献以散。尊者举觯。卑者举角。五献之尊,门外缶,门内壶。君尊瓦甒。

At the sacrifices of the ancestral temple, the highest in rank presented a cup (of spirits to the representative of the dead), and the low, a san (containing five times as much). At some other sacrifices, the honourable took a zhi (containing 3 cups), and the low a horn (containing 4). At the feasts of viscounts and barons, when the vase went round 5 times, outside the door was the earthen ware fou (of supply), and inside, the hu; while the ruler's vase was an earthenware wu.

(《礼记·礼器》,James Legge 译)

原文出现了各种祭祀时使用的盛酒器皿。宗庙的祭祀,地位高的人用容量一升的"爵"敬献。地位低的人用容量五升的"散"敬献。尊贵的人举容量三升的"觯"敬献。卑微的人举容量四升的"角"敬献。子爵、男爵宴饮,最大的盛酒瓦器"缶",放在门外,较小的"壶"放在门内。而君的尊却用最小的"瓦甒"。译文采用了意译、音译、加文内注的方法,传达出器皿的功能、容量、材质等。

三、社会

社会文化负载词反映了中国古代的社会实践、社会结构等,主要包括不同的社会阶层、农业制度等,如《诗经》中"臣工",指周王的群臣百官,可译为minister。再比如:

【例8】

士不偏不党,柔而坚,虚而实。其状朖然不儇,若失其一。傲小物而志属于大,似无勇而未可恐狼,执固横敢而不可辱害,临患涉难而处义不越,南面称寡而不以侈大,今日君民而欲服海外,节物甚高而细利弗赖,耳目遗俗而可与定世,富贵弗就而贫贱弗揭,德行尊理而羞用巧卫,宽裕不訾而中心甚厉,难动以物而必不妄折。此国士之容也。

A scholar-knight is neither partial nor partisan; he is weak yet strong, empty yet full. His manner is transparent, with no suggestion of cunning, as if he were lost in his unity. Oblivious to small matters, his mind is set on great things. He seems cowardly, yet he cannot be frightened. Holding fast obstinately and daring beyond reason, he cannot be threatened with shame or injury. Facing troubles or involved in difficulties, he cleaves to his code of conduct and will not transgress it. When he faces south and calls himself "unworthy," there is no trace of the exaggerated or grandiose. Were he one day to become lord to his people and desire the allegiance of those beyond the seas, he would be economical, very noble, and unconcerned with trivial advantages. In what he does hear and see, he transcends the ordinary, so he can settle the affairs of the world. Wealth and honor he does not pursue, nor does he flee poverty and humble station. His acts of kindness adhere to reason; he would be ashamed to employ artifice or hyperbole. Liberal and generous, he does not revile others, yet he holds the strictest of standards in his heart. It would be difficult to tempt him with material things; he certainly would not rashly compromise his principles. Such is the comportment of a scholar-knight of state.

(《吕氏春秋》,Knoblock、Riegel 译)

"士"是《吕氏春秋》中记述的社会阶层或社会群体之一,当时的社会盛行养"士"之风,这一段论述了"士"的品德。

在汉学家诺布诺克(John Knoblock)和里格尔(Jeffrey Riegel)合译的《吕氏春秋》(*The Annals of Lü Buwei*)中,"士"被译为scholar-knight。该译本的术语表中,译者对scholar-night进行了如下解释:

Scholar-knights 士: Members of a learned class from which rulers of the Warring States period hoped to recruit officials to advise them and serve in heir governments. Scholar-knights were from the lower fringe of the old aristocracy of Western Zhou and Spring and Autumn times—men of good birth but without titles of nobility and hence lacking the opportunity to hold hereditary office. Much of the thought of the *Lüshi chungiu* is derived from the ideology of this class and aims to advise its members on how they should comport and train themselves in order to win the attention and respect of rulers.

【例9】

钧石之石,五权之名,石重百二十斤。后人以一斛为一石,自汉已如此,"饮酒一石不乱"是也。挽蹶弓弩,古人以钧石率之,今人乃以粳米一斛之重为一石,凡石者以九十二斤半为法,乃汉秤三百四十一斤也。

Dan was one of the five units of weight in ancient times and one dan was equal to 120 jin. People of later generations used it to refer to the unit of capacity and one dan was equal to one hu. As an old saying goes, "There are people who are able to drink one dan of wine without getting drunk." Ancient people used jun and dan to measure the strength of a person to draw a bow or a crossbow. Currently people take one hu of polished round-grained rice as one dan, which is about 92.5 jin of today or 341 jin in the Han Dynasty.

(《梦溪笔谈·辩证卷三》,王宏、赵峥 译)

原文记述了重量单位的度量。钧石的石,是重量单位的名称,一石重一百二十斤。后人把一斛作为一石,在汉代已经如此,所谓"饮酒一石不乱"就是。开弓张弩,古人用钧石来计算,现在人以一斛粳米的重量为一石,这种石相当于九十二斤半,就是汉代的三百四十一斤。重量单位均采用了音译。

四、宗教

宗教文化负载词反映了中国古代的神话传说、宗教信仰等,主要包括神话人物、祭祀仪式等。比如:

【例10】

凡南次二经之首,自柜山至于漆吴之山,凡十七山,七千二百里。其神状皆龙身而鸟首。其祠:毛用一璧瘗,糈用稌。

With Mount Guishan being the first mountain of the second mountain range of the Southern Mountains and Mount Qiwu being the last, there are altogether seventeen mountains, which cover a total distance of 7,200 li. The deities of these mountains all have a dragon's body and a bird's head. The ceremony of offering sacrifices to them goes as follows: 1. Bury the sacrificial animals together with a jade disc; 2. Use sticky rice as the sacrificial rice.

（《山海经·南山经》，王宏、赵峥 译）

《山海经》中记述了诸多奇形怪状的山神，或是半人半兽，或是各种动物形体的组合，体现了原始宗教自然崇拜的特色。译文将山神的奇异外形和祭拜方式直译出来。

【例11】

是月也，大饮蒸，天子乃祈来年于天宗。大割，祠于公社及门闾，飨先祖五祀，劳农夫以休息之。天子乃命将率讲武，肄射御、角力。

In this month, a large drinking party is to be held after steamed sacrifices are offered to gods. The king is to pray to the heavenly gods for a good harvest in the coming year. There will be a large-scale slaughter of animals to be offered to the shrines, doors and ancestors in the five kinds of sacrificial ceremonies and to be given to the peasants so that they can rest. The king is to order the military officers to practise military skills, including archery and the driving of chariots, and to hold matches of strength.

（《吕氏春秋》，汤博文 译）

原文记述的是中国古代的祭祀仪式。“大割”指杀割群牲以祭祀，直译为a large-scale slaughter of animals.“公社”和“门闾”分别是官家和私人祭祀的场所，意译为the shrines, doors and ancestors.“五祀”指祭祀门、户、中霤、灶、行五种神祇，直译为the five kinds of sacrificial ceremonies.

【例12】

菜品中芜菁、菘、芥之类，遇旱其标多结成花，如莲花，或作龙蛇之形。此常性，无足怪者。熙宁中，李宾客及之知润州，园中菜花悉成荷花，仍各有一佛坐于花中，形如雕刻，莫知其数。曝干之，其相依然。或云：“李君之家奉佛甚笃，因有此异。”

In dry days, vegetables such as turnips, cabbages and leaf mustards are likely to change into flowers resembling lotuses or dragons and snakes. Such changes are very common. There is no need to be surprised. During Xining period of the reign of Emperor Shenzong, Li Jizhi, the subordinating official of the Crown Prince, was the prefect of Runzhou. In his vegetable garden, all the vegetables turned into flowers resembling lotuses. Inside each flower was something similar to the statue of Buddha. There were many such statues of Buddha whose shapes remained intact even when the flowers were dried in the sunlight. People said, "The Li family is pious to Buddhism. This explains why the miraculous phenomenon takes place in his house."

(《梦溪笔谈·神奇卷二十》,王宏、赵峥 译)

五、语言

语言文化负载词主要包括与农业、农耕相关的谚语和俗语,如"亡羊补牢,未为晚也"(It is not too late to mend the ranch after you lost some sheep.),"男耕女织"(Men till; women weave.),"三折臂知为良医"(He whose arm has broken twice will make a good surgeon.),"子欲富,黄金覆"(Golden banking makes you rich.)。再比如:

【例13】

此物性不耐寒。阳中之树,冬须草裹;不裹即死。其生小阴中者,少禀寒气,则不用裹。所谓"<u>习以性成</u>"。一木之性,寒暑异容;若朱、蓝之染,能不易质?故<u>观邻识士,见友知人</u>也。

This plant is not cold-resistant. Those individuals formerly grown in sunny spots must be wrapped with straw in winter, or else they will be killed by the cold. Those formerly grown under mild shade have, however, acquired cold-tolerance early in life and need no wrappings. Thus "habit makes nature". The same shrub differs in capacity to withstand cold by usage. Just as contact with ochre and indigo alters the color, natural disposition will change with the environment. Therefore "discern a person by watching his neighbours and judge a man by studying his friends".

（《齐民要术》卷四《种椒第四十三》，石声汉 译）

（花椒）这植物不耐寒：原来长在阳地的树，冬天须要用草包裹；不包裹就会冻死。生在比较上向阴处所的，从小获得了寒冷的习惯，就不必包裹。这就是所谓"习惯成本性"。一种树的本性，耐寒与否，有不同的表现；正像碰着红土蓝淀，就会染上颜色一样，性质怎能不发生改变？所以由邻居和朋友，可以推想到某人的性情和为人。这一段说明我国古人对于环境因素影响遗传变异的认识，同时蕴含了"近朱者赤，近墨者黑"的哲理。

【例14】

"甘受和，白受采。"世间丝、麻、裘、褐皆具素质，而使殊颜异色得以尚焉。

"The bland will absorb a mixture of tastes, and the white will absorb the rainbow colors." All the silk, hemp, fur and woolen stuffs are naturally plain, so they can be dyed different colors which make them valuable.

（《天工开物·彰施第三》，王义静、王海燕、刘迎春 译）

练　习

1. 翻译以下典籍原文，特别注意文化负载词的翻译方法。

1）小豆，忌卯；稻、麻，忌辰；禾，忌丙；黍，忌丑；秫，忌寅、未；小麦，忌戌；大麦，忌子；大豆，忌申、卯。凡九谷有忌日；种之不避其忌，则多伤败。此非

虚语也！其自然者,烧黍穰则害瓠。(《氾胜之书》)

2)小麦,忌戌,大麦,忌子。除日不中种。(《氾胜之书》)

3)天子龙衮,诸侯黼,大夫黻,士玄衣纁裳;天子之冕,朱绿藻十有二旒,诸侯九,上大夫七,下大夫五,士三。此以文为贵也。(《礼记》)

4)致中和,天地位焉,万物育焉。(《中庸》)

5)使天下之人齐明盛服,以承祭祀,洋洋乎如在其上,如在其左右。(《中庸》)

6)耧耩稀种,一斗可种一亩;量家田所需种子多少,而种之。(《齐民要术》)

7)谚曰:"湿耕泽锄,不如归去。"(《齐民要术》)

8)谚曰:"耕而不劳,不如作暴。"(《齐民要术》)

9)凡栽树,正月为上时,谚曰:"正月可栽大树。"言得时则易生也。二月为中时,三月为下时。(《齐民要术》)

10)天孙机杼,传巧人间。(《天工开物》)

11)凡南次三经之首,自天虞之山以至南禺之山,凡一十四山,六千五百三十里。其神皆龙身而人面。其祠:皆一白狗祈,糈用稌。(《山海经》)

2.英译以下句子及段落。

1)十天干即甲、乙、丙、丁、戊、己、庚、辛、壬、癸。

2)十二地支即子、丑、寅、卯、辰、巳、午、未、申、酉、戌、亥。

3)二十八宿指二十八颗星,即东方的角、亢、氐、房、心、尾、箕;南方的井、鬼、柳、星、张、翼、轸;西方的奎、娄、胃、昴、毕、觜、参;北方的斗、牛、女、虚、危、室、壁。

4)耒耜是中国古代一个生活在黄河中下游的部落首领神农发明的农具,它是先秦时期的主要农耕工具,主要用于农业生产中的翻整土地、播种庄稼。耒耜的发明提高了耕作效率,使人们的劳动强度大大减轻,也使谷物产量大大增加。河南辉县还出土过战国时期的铁犁铧。铁犁铧的发明又是一个了不起的成就,它标志着人类改造自然的斗争进入一个新的阶段。

5)祭祀是一种信仰活动,源于天地和谐共生的信仰理念。据现代人类学、考古学的研究成果表明,人类最原始的两种信仰:一是天地信仰,二是祖先信仰。天地信仰和祖先信仰产生于人类初期对自然界以及祖先的崇拜,由此产生了各种崇拜祭祀活动。

6)五祀是中国古代一项重要的祭祀,起源自先秦,延续至明清,时至今日在我国不少地方仍可见其遗俗。先秦时期的五祀是指对户、灶、中霤、门、行五种小神之祀。

7)《吕氏春秋》成书于秦统一中国前夕,由秦相吕不韦组织编纂而成。按照司马迁《史记》的说法,吕不韦仿效战国四公子的做法招养门客三千,让他们把自己所闻所见和感想都写出来而汇集成书,取名《吕氏春秋》。书中内容以儒、道思想为主,亦取墨、法、名、农、阴阳等诸家学说,总结了春秋战国时百家争鸣的成果,为秦统一天下服务。《吕氏春秋》语言简洁、生动、形象,说理极富逻辑性。

8)农耕文化,即古代农业文化,指在中国原始农业和传统农业时期,通过生产实践活动所创造出来的与农业有关的物质文化和精神文化的总和,包括农业科技、思想、制度、习俗、饮食等。在中国文化产生和发展的过程中发挥着基础作用,影响着中华民族的生存方式,塑造着中华民族文化的自身。中原农耕文化是中国农耕文化的重要发源地之一,是宋代以前中国农业文化的轴心。河南裴李岗文化出土的农业生产工具为早期农耕文化的发达提供了实物证据。

3. 以下三段讲述了三个关于“士容”的小故事,根据释文进行翻译。

齐有善相狗者,其邻假以买取鼠之狗,期年乃得之,曰:“是良狗也。”其邻畜之数年,而不取鼠,以告相者。相者曰:“此良狗也。其志在獐麋豕鹿,不在鼠。欲其取鼠也则桎之。”其邻桎其后足,狗乃取鼠。夫骥骜之气,鸿鹄之志,有谕乎人心者,诚也。人亦然。诚有之,则神应乎人矣,言岂足以谕之哉？此谓不言之言也。

客有见田骈者，被服中法，进退中度，趋翔闲雅，辞令逊敏。田骈听之毕而辞之。客出，田骈送之以目。弟子谓田骈曰："客，士欤?"田骈曰："殆乎非士也。今者客所弇敛，士所术施也，士所弇敛，客所术施也。客殆乎非士也。"故火烛一隅，则室偏无光；骨节蚤成，空窍哭历，身必不长；众无谋方，乞谨视见，多故不良；志必不公，不能立功；好得恶予，国虽大不为王；祸灾日至。故君子之容，纯乎其若钟山之玉，桔乎其若陵上之木。淳淳乎慎谨畏化，而不肯自足；乾乾乎取舍不悦，而心甚素朴。

唐尚敌年为史，其故人谓唐尚愿之，以谓唐尚。唐尚曰："吾非不得为史也，羞而不为也。"其故人不信也。及魏围邯郸，唐尚说惠王而解之围，以与伯阳，其故人乃信其羞为史也。居有间，其故人为其兄请。唐尚曰："卫君死，吾将汝兄以代之。"其故人反兴再拜而信之。夫可信而不信，不可信而信，此愚者之患也。知人情，不能自遗，以此为君，虽有天下何益？故败莫大于愚。愚之患，在必自用。自用则戆陋之人从而贺之。有国若此，不若无有。古之与贤，从此生矣。非恶其子孙也，非徼而矜其名也，反其实也。

(《吕氏春秋》)

【释文】

齐国有个擅长相狗的人，邻居委托他买一只捕捉老鼠的狗。他用了整整一年时间才买到，对邻居说："这是一只好狗。"他的邻居喂养了好几年，狗却不捕捉老鼠，邻居就把这一情况告诉了相狗的人。相狗的人说："这是一只好狗。它志在猎取獐、麋、猪、鹿，不在捉老鼠。想让它捕捉老鼠就得把它桎梏起来。"邻居桎梏了狗的后腿，狗才捉老鼠。骥骜的气质，鸿鹄的心志，能在人的心灵中被感悟到，是因为这种气质和心志的精诚。人也是如此。具备如此精诚的气质和心志，就会在人的精神风貌上反映出来了，语言哪能足以表达呢？这叫无须用言语表述的语言。

有个客人前来拜见田骈，他的服饰合乎法式，进退合乎礼仪，行止娴静优雅，言辞谦逊敏捷。田骈听他说完就谢绝了他。客人离去的时候，田骈目送他出门。弟子们对田骈说："这位客人是个士人吗？"田骈说："恐怕不是士人吧。刚才来客掩饰、收敛的地方，正是士人狂放不羁之处；而士人所掩饰、收敛的地方，正是客人狂放不羁之处。这位客人大概不是个士人吧。"所以说，火光只照一个角落，房间的另一半就没有光亮。骨骼过早长成，空窍就会疏松不实，身材一定长不高大。众人不讲求道义，只局限于小心整饬自己的外表，就会巧诈多端、心思不正。心志不正，就不能建立功业。喜欢聚敛而厌恶施舍，国家再大也不能称王天下，灾祸还会天天发生。所以，君子的仪容，像昆仑山的玉石一样美好，像高山上的大树一样伟岸。他们纯洁朴实，言行审慎，警惕事物的发展变化，而不会骄傲自满；他们孜孜不倦，取舍严肃而不苟且，而且心地非常淳朴。

唐尚刚到了做史官的年龄，他的朋友以为唐尚想做史官，就跟唐尚提起这件事。唐尚说："我并不是不能够做史官，而是羞于做这样的官。"朋友不相信。等到魏国围困邯郸的时候，唐尚游说魏惠王从而解了邯郸之围，赵国把伯阳封给唐尚。他的朋友这才相信他真的是羞于做史官。过了一段时间，这个朋友请求唐尚帮他的哥哥谋求一官半职。唐尚说："卫国君主死后，我就让你哥哥取代他的位置。"他的朋友反倒站起身来拜了两次，对唐尚的话信以为真。对可信的不相信，对不可信的反倒相信，这是愚蠢的人的通病。知道人之常情，却不能按照它反躬自省，这样做君主，即使拥有了天下，又有什么用？所以，没有比愚蠢更能败坏事情的了。愚蠢的人的弊病，在于刚愎自用。君主刚愎自用，鄙陋无知的人就会趋从并且祝贺他。这样占有国家，就不如没有。古代把天下让给贤人，就是由此产生的。让贤并不是出于对自己子孙的憎恶，也不是追求或者炫耀虚名，而是为了返回生命的本性。

延伸阅读

《吕氏春秋》译文赏析

《吕氏春秋》共有三个英译本，一个是诺布诺克和里格尔的合译本(以下简称“合译”)，一个是翟江月译本(以下简称“翟译”)，一个是汤博文译本(以下简称“汤译”)。

下文选自《吕氏春秋·有始览第一·有始》，文中对生态文化负载词“九野”“九州”“九山”“九塞”“九薮”“八风”“六川”进行了详细解释。

何谓九野？中央曰钧天，其星角、亢、氐；东方曰苍天，其星房、心、尾；东北曰变天，其星箕、斗、牵牛；北方曰玄天，其星婺女、虚、危、营室；西北曰幽天，其星东壁、奎、娄；西方曰颢天，其星胃、昴、毕；西南曰朱天，其星觜嶲、参、东井；南方曰炎天，其星舆鬼、柳、七星；东南曰阳天，其星张、翼、轸。

合译：What are called the “nine fields”? The very center is called the Hub of Heaven, comprising the zodiac signs Horn, Neck, and Root. Due east is called Azure Heaven, comprising the zodiac signs Room, Heart, and Tail. The northeast is called Changing Heaven, comprising the zodiac signs Winnowing Basket, Dipper, and Herdboy. Due north is called Dark Heaven, comprising the zodiac signs Serving Maid, Emptiness, Rooftop, and Encampment. The northwest is called Gloomy Heaven, comprising the zodiac signs Eastern Wall, Legs, and Bond. Due west is called Luminous Heaven, comprising the zodiac signs Stomach, Pleiades, and Net. The southwest is called Vermilion Heaven, comprising the zodiac signs Turtle, Triad, and Eastern Well. Due south is called Fiery Heaven, comprising the zodiac signs Carriage Ghost, Willow, and Seven Stars. The southeast is called Yang Heaven, comprising the zodiac signs Extended Net, Wings, and Chariot Platform.

翟译：What are these nine parts of Heaven? The central part is called "the even sky", and the important stars of this part are Jue (α of Virgo), Kang (a group of four stars also belonging to Virgo) and Di (a group of four stars, including α, ι, γ and β of Libra). The eastern part is called "the blue sky", and the important stars of this part are Fang (a group of four stars, including π, ρ, δ and β of Scorpio), Xin (a group of three stars, including σ, α and τ of Scorpio) and Wei (a group of nine stars also belonging to Scorpio). The northeast part is called "the ever-changing sky", and the important stars of this part are Ji (a group of four stars belonging to Sagittarius), Dou (a group of six stars also belonging to Sagittarius) and Qian Niu (a group of stars belonging to Capricornus). The northern part is called "the black sky", and the important stars of this part are Wu Nü (a group of stars belonging to Water Bearer), Xu (a group of stars including β of Water Bearer and α of Equuleus), Wei (a group of three stars, including α of Water Bearer, θ and ε of Pegasus) and Ying Shi (a group of two stars, including α and β of Pegasus). The northwest part is called "the dark sky", and the important stars of this part are Dong Bi (a group of three stars, including γ of Pegasus and α of Andromeda), Kui (a group of sixteen stars, nine of which belong to Andromeda and seven of which belong to Pisces) and Lou (a group of three stars—α, β and γ of Aries). The western part is called "the bright sky", and the important stars of this part are Wei (a group of three stars belonging to Aries), Mao (a group of eight stars belonging to Taurus) and Bi (a group of stars also belonging to Taurus). The southwest part is called "the red sky" and the important stars of this part are Zi Xi (a group of three stars belonging to Taurus), Shen (a group of seven stars belonging to Orion) and Dong Jing (a group of eight stars belonging to Gemini). The southern part is called "the hot sky", and the important stars of this part are Yu Gui (a group of four stars belonging to Cancer), Liu (a group of eight stars belonging to Hydra) and the Seven Stars (a group of stars belonging to Capricornus). The southeast part is called "the sunny sky", and the important stars of this part are Zhang (a group of six stars belonging to Hydra), Yi (a group of twenty-two stars belonging to Hydra and Crater) and Zhen (a group of four stars, including β, γ, δ and ε of Corvus).

汤译：Which are the nine segments? The central segment is called the Equilibrating Sky, where the constellations Jiao, Kang and Di are. The eastern segment is called the Blue Sky, where the constellations Fang, Xin and Wui are. The northeastern segment is called the Changing Sky, where the constellations Qi, Dou and Qianniü are. The northern segment is called the Black Sky, where the constellations Wuniü, Xu, Wei and Yingshi are. The northwestern segment is called the Dark Sky, where the constellations Dongbi, Kui and Lou are. The western segment is called the White Sky, where the constellations Wuei, Ang and Bi are. The southwestern segment is called the Red Sky, where the constellations Zixi, Can and Dongjing are. The southern segment is called the Fiery Sky, where the constellations Gui, Liu and Seven Stars are. The southeastern segment is call the Sunny Sky, where the constellations Zhang, Yi and Zhen are.

何谓九州？河、汉之间为豫州，周也。两河之间为冀州，晋也。河、济之间为兖州，卫也。东方为青州，齐也。泗上为徐州，鲁也。东南为扬州，越也。南方为荆州，楚也。西方为雍州，秦也。北方为幽州，燕也。

合译：What are called the "nine provinces"? Between the Yellow and Han rivers is Yu Province, corresponding to Zhou. Between the two parts of the Yellow river is Ji Province, corresponding to the state of Jin. Between the Yellow and Ji rivers is Yan Province, corresponding to the state of Wey. Due east is Qing

Province, corresponding to the state of Qi. Beyond the Si River is Xu Province, corresponding to the state of Lu. In the southeast is Yang Province, corresponding to the state of Yue. Due south is Jing Province, corresponding to the state of Chu. Due west is Yong Province, corresponding to the state of Qin. Due north is You Province, corresponding to the state of Yan.

翟译：What are these nine prefectures? The area between the Yellow River and the Han River is called "Yu Zhou", and it belongs to the royal family of the Zhou Dynasty. The area between the Qing River and the Xi River is called "Ji Zhou", and it belongs to the state of Jin. The area between the Yellow River and the Ji River is called "Yan Zhou", and it belongs to the state of Wei. The prefecture located in the east is called "Qing Zhou", and it belongs to the state of Qi. The prefecture south of the Si River is called "Xu Zhou", and it belongs to the state of Lu. The prefecture located in the southeast is called "Yang Zhou", and it belongs to the state of Yue. The prefecture located in the south is called "Jing Zhou", and it belongs to the state of Chu. The prefecture located in the west is called "Yong Zhou", and it belongs to the state of Qin. The prefecture located in the north is called "You Zhou", and it belongs to the state of Yan.

汤译：Which are the nine prefectures? The prefecture Yu between the Yellow and Hanshui rivers is the territory of the Zhou dynasty. The prefecture Ji between the Clear and Yellow rivers is the territory of the state of Jin. The prefecture Yan between the Yellow and Jishui rivers is the territory of the state of Wei. The prefecture Qing in the east is the territory of the state of Qi. The prefecture Xu by the Sishui River is the territory of the state of Lu. The prefecture Yang in the southeast is the territory of the state of Yue. The prefecture Jing in the south is the territory of the state of Chu. The prefecture

Yong in the west is the territory of the state of Qin. The prefecture You in the north is the territory of the state of Yan.

何谓九山? 会稽,太山,王屋,首山,太华,岐山,太行,羊肠,孟门。

合译: What are called the "nine mountains"? They are Kuaiji, Mount Tai, Wangwu, Mount Shou, Taihua, Mount Qi, Taihang, Yangchang, and Mengmen.

翟译: What are these nine huge mountains? They are Kuai Ji Mountain, Tai Mountain, Wang Wu Mountain, Shou Yang Mountain, Tai Hua Mountain, Qi Mountain, Tai Hang Mountain, Yang Chang Mountain and Meng Men Mountain.

汤译: Which are the nine mountains? They are the Kuaiji, Taishan, Wangwu, Shoushan, Taihua, Qishan, Taihang, Yangchang and Mengmen mountains.

何谓九塞? 大汾,冥阨,荆阮,方城,殽,井陉,令疵,句注,居庸。

合译: What are called the "nine passes"? They are Dafen, Ming'e, Jingruan, Fangcheng, Yao, Jingxing, Lingci, Gouzhu, and Juyong.

翟译: What are these nine natural strategic passes? They are Da Fen, Ming Ai, Jing Ruan, Fang Cheng, Xiao, Jing Xing, Ling Ci, Gou Zhu and Ju Yong.

汤译: Which are the nine passes? They are the Dafen, Minge, Jingyuan, Fangcheng, Xiao, Jingjing, Lingpi, Gouzhu and Juyong passes.

何谓九薮? 吴之具区,楚之云梦,秦之阳华,晋之大陆,梁之圃田,宋之孟诸,齐之海隅,赵之钜鹿,燕之大昭。

合译: What are called the "nine marshes"? They are Juqu of Wu, Yunmeng of Chu, Yanghua of Qin, Dalu of Jin, Putian of Liang, Mengzhu of Song, Haiyu of Qi, Julu of Zhao, and Dazhao of Yan.

翟译: What are these nine great lakes? They are the Ju Qu Lake of the state of Wu, the Yun Meng Lake of the state of Chu, the Yang Hua Lake of the state of Qin, the Da Lu Lake of the state of Jin, the Pu Tian Lake of the state of Liang, the Meng Zhu Lake of the state of Song, the Hai Yu Lake of the state of Qi, the Ju Lu Lake of the state of Zhao and the Da Zhao Lake of the state of Yan.

汤译：Which are the nine great lakes? They are lakes Mengyun in Chu, Yanghua in Qin, Dalu in Jin, Putian in Liang, Mengzhu in Song, Haiyu in Qi, Julu in Zhao and Dazhao in Yan.

何谓八风？东北曰炎风，东方曰滔风，东南曰熏风，南方曰巨风，西南曰凄风，西方曰飂风，西北曰厉风，北方曰寒风。

合译：What are called the "eight winds"? The northeast is called Fiery Wind, due east Torrent Wind, southeast Smoky Wind, due south Giant Wind, southwest Chilling Wind, due west Whirlwind, northwest Sharp Wind, and due north Cold Wind.

翟译：What are the eight different kinds of winds? The northeasterly is called Yan Feng (which means very hot wind). The easterly is called Tao Feng (which means strong wind originating from the sea). The southeasterly is called Xun Feng (which means gentle wind). The southerly is called Ju Feng (which means very heavy wind). The southwesterly is called Qi Feng (which means very sharp and unmerciful wind). The westerly is called Liu Feng (which means harmfully heavy and sharp wind). The northwesterly is called Li Feng (which means bitterly sharp wind). And the northerly is called Han Feng (which means fiercely chilly wind).

汤译：Which are the eight types of wind? Wind from the northeast is called Hot wind; from the east, Billowy wind; from the southeast, Warm wind; from the south, Great wind; from the southwest, Sad wind; from the west, High wind; from the northwest, Fierce wind; and from the north Cold wind.

何谓六川？河水，赤水，辽水，黑水，江水，淮水。

合译：What are called the “six rivers”? The Yellow River, the Chi River, the Liao River, the Hei River, the Yangzi River, and the Huai River.

翟译：What are these six big rivers? They are the Yellow River, the Red River, the Liao River, the Black River, theYangtze River and the Huai River.

汤译：Which are the six great rivers? They are the Yellow, Red, Liao, Black, Changjiang and Huai rivers.

第三节 数量词的翻译

数词和量词在农业科技典籍中出现的频率很高，用来表达时间、长度、面积、体积等。与普通科技文本不同的是，典籍中的数词和量词是按照古代的方式计量，并有古代汉语的特殊用法，如“年方二八”是指十六岁，而不是二十八岁；“虽九死其犹未悔”中的“九”泛指多，而不是死了九次。在翻译时，必须正确理解数量词的具体含义，选择合适的翻译策略和方法传达信息。

一、数字

数量有确数和概数之分。顾名思义，确数指具体的、准确的数量，概数指大约估计的数量。这两种数量在典籍中都能用数字来表达，翻译时大多可以采取直译，直接对应英语中的数字。但有些表达概数的数字，如上文中提到的“九”，需要意译，英译文中往往不会出现对应的数字。

【例1】

……此十四种，早熟、耐旱、免虫；……二种味美。……此二十四种，穗皆有毛，耐风，免雀暴；……一种易舂；……此三十八种中……二种味美；……三种味恶；……二种易舂。……此十种晚熟，耐水；有虫灾则尽矣。

These 14 sorts ripen early, are drought-resistant, and immune from insect-pests, ... 2 of them are delicious. ... Of these 24, ears are setigerous, wind-resistant, immune from sparrows; ... 1 easily pounded; ... of these 38, ... 2 are delicious; ... 3 have an unpleasant flavour; ... 2 are easily pounded. ... Of these 10 ripen late, are flood-resistant, but liable to destruction by insects.

（《齐民要术》卷一《种谷第三》，石声汉 译）

贾思勰在《齐民要术》中记载了86个品种的粟，并根据其质性和米味进行了分类，列举中不乏数字的使用。原文中的数字是确定的数量，均直译为对应的阿拉伯数字。

【例2】

（合手药法：）白桃人二七枚，去黄皮，研碎，酒解取其汁。

(Hand ointment:) Take 27 white peach kernels, peel off the brown skin, triturate, steep in some wine, and drain to obtain a clear liquid.

（《齐民要术》卷五《种红蓝花及栀子第五十二》，石声汉 译）

【例3】

盖覆器口，安硎苦耕反、泉、冷水中，使冷气折其出势。得三七日，然后剖生；养之，谓为"爱珍"，亦呼"爱子"。绩成茧，出蛾，生卵；卵七日又剖成蚕；多养之，此则"爱蚕"也。

Cover the mouth, and immerse the belly of the vessel under cold water of a stream, ravine or fountain to delay the hatching down to 3 weeks' time through cold treatment. These later hatching caterpillars are aichen (or ai-tzu) which spin, metamorphose and spawn in the regular course of time. Ova of ai-chen hatch in one week or so as usual. Caterpillars from these ova are called "ai", and are usually again kept in large numbers.

(《永嘉记》,石声汉 译)

以上两例中的"二七枚"和"三七日"虽然形式上相同,代表的数字却不一样。"二七枚"指27枚,与前例中的"二十四种"和"三十八种"不同,这里的两位数中间没有数字"十"。"三七日"是三个七日,指21天,即三周时间。

【例4】

(小豆:)豆角三青两黄,拔而倒竖笼丛之,生者均熟。

(For Lesser beans:) When 3/5 of the pods are still green, pull out, stand the bunches upside down bound in bundles, then the green pods will gradually all ripen.

(《齐民要术》卷二《小豆第七》,石声汉 译)

"三青两黄"是指豆荚大半青小半黄的时候,此处用分数进行直译,"青"和"黄"的占比一目了然。

【例5】

(胡麻:)以五六束为一丛,斜倚之。候口开,乘车诣田斗薮,倒竖,以小杖微打之。还从之。三日一打,四五遍乃尽耳。

(For sesame): Make bundles of 5 or 6 bunches, and slant them so that they interlock. When the follicles dehisce, ride on a cart through the field, tap the bundles with a small baton and collect the ripe seeds. Tap every three days. After 4 or 5 tappings, no more will be left (on the stalk).

(《齐民要术》卷二《胡麻第十三》,石声汉 译)

原文记述的是胡麻(即芝麻)的收获方法,文中数字也采用了直译的处理方法。试比较:这里的"四五"和下一例中的"四五"意思是否相同?

【例6】

(区种瓜:)又,可种小豆于瓜中,亩四五升;其藿可卖。

(Shallow-pit cultivation of melons:) Or lesser beans, 4 to 5 *shêng* per mu, may be sown among the melons; the bean-leaves can be sold as greens.

(《氾胜之书》,石声汉 译)

直译的前提是准确理解数字的含义。同样是"四五",意思却不尽相同。前者是"四或五"的选择,表达确数;后者是"四至五"的范围,表达概数。

【例7】

七月中,作坑,令受百许束。

In the middle of the 7th month, dig a pit big enough to receive about 100 bunches of indigo plants.

(《齐民要术》卷五《种蓝第五十三》,石声汉 译)

【例8】

(瓜蔓:)有蚁者,以牛羊骨带髓者,置瓜科左右;待蚁附,将弃之。弃二三,则无蚁矣。

(For melon-vines:) If plagued by ants, place a few pieces of marrowy ox or sheep bones by the melon stock. Wait till ants assemble. Pick up and cast both away. Repeating thus several times will clear away all the ants.

(《齐民要术》卷二《种瓜第十四》,石声汉 译)

以上两例中的"百许"和"二三"都表达概数,但翻译时采用了不同的方法。"百许"的意思是一百左右,直译为 about 100。"二三"是指几次,不是确定的次数,因此意译为 several times,译文中没有出现对应的数字 two 和 three。

【例9】

凡五谷地畔近道者,多为六畜所犯;宜种胡麻、麻子以遮之。胡麻,六畜不食;麻子啮头则科大。收此二实,足供美烛之费也。

Cereal - fields abutting on thoroughfares are usually violated by passing animals. Have sesame or female hemp plants on the boundary. Cattle will never touch sesame; female hemp when browsed upon branches only more profusely. Both plants yield oily seeds good for illumination.

(《齐民要术》卷二《种麻子第九》,石声汉 译)

【例10】

一人治之,十人食之,六畜皆在其中矣。

This formula of one person working on the farm to feed ten persons includes the feeding of animals.

(《吕氏春秋》,汤博文 译)

文中数字"五"和"六"都不是具体的数量,"五谷"泛指粮食类作物,意译为cereal;"六畜"泛指动物、牲口,意译为animals,cattle。源自农耕文化的祝福语"五谷丰登"、"六畜兴旺"等都是泛指粮食丰收、牲畜家禽繁衍兴盛。再如《管子·地员篇》中的"五沃之土"泛指土质肥沃的上等土壤。同样,下一例中的"五果"也泛指各种果树。

【例11】

凡五果,花盛时遭霜,则无子。

If a frost occurs when fruit-trees are in blossom, no fruit will be formed.

(《齐民要术》卷四《栽树第三十二》,石声汉 译)

二、时间

《齐民要术》记述的"合香泽法"中有这样一句:夏一宿,春秋再宿,冬三宿。意思是:夏天,过一夜;春天秋天,两夜;冬天,三夜。石声汉将这句话译为:Overnight in summer, 2 nights in spring and autumn, 3 nights in winter. "一宿”没有英译为对应的数字，而是用了 overnight；“再宿”本身没有数字，英译时用了数字 2。再如前面提到的“三七日”，是指 21 天。农耕文化造就了典籍中特殊的时间表达方式，翻译时要注意转换成现代时间表达法，必要时应加注解释。

【例 12】

冬至后五旬七日，昌生。昌者，百草之先生也。于是始耕。

Fifty-seven days after the Winter Solstice, calamus will begin to sprout. Among all the weeds, calamus is the first to sprout. It is time to begin to plow the fields.

（《吕氏春秋》，汤博文 译）

“旬”指的是十天，“五旬”即 50 天，“五旬七日”译为 fifty-seven days。“昌”是菖蒲，一年中最早出现的宿根草本植物。冬至后五十七天，菖蒲出现，就可以开始耕田了。

本章开头“煮胶法”一例中有这样一句：煮胶要用二月、三月、九月、十月。译文为：Glue-making ought to be done in the 2nd, 3rd, 9th and 10th months (about March, April, October and November). 中国古代使用的是农历纪月法，与英语中公历纪月法不同，在表述上也并非一一对应，因此句中的月份没有译为 February, March, September, October，而是用序数词译为 the 2nd, 3rd, 9th and 10th months，然后加注说明对应的公历月份，帮助英语读者理解，有利于中国农耕文化的传播。类似的例子还有：

【例 13】

六月，作苇屋覆之。不耐寒热顾也。九月掘出置屋中。中国多寒，宜作窖，以谷稈合埋之。

In the 6th month, make a canopy over them with a thatch of rushes. The hot summer is too much for them. In the 9th month, dig them out and remove under a roof—the northern China winter is much too severe. It is best to store them in dugouts with chaff all over to keep warm.

(《齐民要术》卷三《种姜第二十七》,石声汉 译)

原文记述的是姜的种植和贮存方法。"六月"和"九月"分别译为the 6th month和the 9th month。文中的"中国"是指黄河流域。

【例14】

梨熟时,全埋之。经年。至春,地释,分栽之;多著熟粪及水。至东,叶落,附地刈杀之,以炭火烧头。二年即结子。

When fruits ripen, bury whole pears into the ground. After one year's growth, transplant during a thaw in the spring. Tender with good dressings of ripened compost and water well. Cut down the stem flush with the ground level and cinder the stump. It will bear good fruits two years later.

(《齐民要术》卷四《插梨第三十七》,石声汉 译)

除了"年",典籍中也常用"岁"字表达年这一时间概念。

【例15】

田,二岁不起稼,则一岁休之。

If a field gave a poor crop in the second year, fallow it for one year.

(《汜胜之书》,石声汉 译)

【例16】

凡开荒山泽田,皆七月芟艾之。草干,即放火。至春而开。根朽省功。其林木大者,㓗杀之;叶死不扇,便任耕种。三岁后,根枯茎朽,以火烧之。入地尽也。

To reclaim any hilly or swampy waste, first cut down the wild vegetations in the 7th month. When the rubbish has dried a little, set fire to it. Plough to open up next spring—the roots are now rotten, so the ground will be easily work. If there be any woody growth too big to be hewn away, girdle it to kill. When its leaves give no more canopying effect, the land will be ready for ploughing and sowing. After 3 years' time, set fire to the withered and rotten trunks, then all these wild growth can eventually be purged down beneath.

（《齐民要术》卷一《耕田第一》，石声汉 译）

"岁"指年，"三岁"即为three years。劄(yīng)的意思是用刀切割树皮。在开荒时要"杀"树而不砍伐，便利用"环割"，切断韧皮部和新木质部，中断树干内营养物质的输送，使茎干自行死亡。①

【例17】

李性坚实晚，五岁始子，是以借栽。栽者三岁便结子也。

Gages are hard by nature and bear fruits rather late, i.e. not until they are 5 years old; therefore it is better to use cuttings. Cuttings usually fruit in 3 years time.

（《齐民要术》卷四《种李第三十五》，石声汉 译）

三、量词

典籍中的有些量词是汉语所特有的，如重量单位"两"、长度单位"尺"、面积单位"亩"等。有些则经过历史演变，在语义和度量上有别于现代汉语中的量词，如容积单位"升"，《氾胜之书》中的1升相当于如今的167毫升。

---

① 贾思勰．齐民要术[M]．石声汉，译．北京：中华书局，2015.

Appendix I. TABLE OF APPROXIMATE EQUIVALENTS OF MEASURES USED IN FAN'S BOOK

(Chinese measures 2,000 years ago.)

1. Length

1 寸(ts'un) = 22 mm.

1 尺(ch'ih)=10 寸=22 cm.

1 丈(chang) = 10 尺= 100 寸= 2.2 m.

2. Volume

1 升(sheng) = 167 mL.

1 斗(tou)=10 升=1.67 L.

1 石(shih)=(=1 斛(hu))=10 斗=100 升= 16.7 L

3. Area

1 畝(mu) =6.67 are.

4. Weight

1 斤(chin) = 177.8 gram

1 石(shih)=120 斤=21.336 kg.

石声汉译释《氾胜之书今释》中的量词表

【例 18】

九月十月中,于墙南日阳中,掘作坑;深四五尺。取杂菜,种别布之,一行菜,一行土。去坎一尺许,便止;以穰厚覆之。得经冬。须即取,粲然与夏菜不殊。

In the 9th or 10th month (late October to early December) dig trenches 4 to 5 *ch'ih* deep at the sunny southern side beneath a wall. Into such trenches pack fresh vegetables in layers (better only one sort in one layer) alternating with layers of earth, until about 1 *ch'ih* below the ground. Then cover with earth and finally with straw. The stored vegetables will last for the winter. Pick out whatever you like whenever wanted, they will be as fresh and crispy as summer-grown ones.

（《齐民要术》卷九《作菹、藏生菜法第八十八》，石声汉 译）

原文记述的是新鲜蔬菜的保藏方法。“九月十月”译为 the 9th or 10th month，并加注说明是公历十月底至十二月初。量词“尺”采用了威氏拼音法音译。“一行菜，一行土”实际上是“多行菜，多行土”，“行”译成了复数形式 layers。

【例 19】

种桑法：五月，取椹著水中，即以手溃之。以水灌洗，取子，阴干。治肥田十亩，荒田久不耕者尤善！好耕治之。每亩，以黍椹子各三升合种之。黍桑当俱生。锄之。桑令稀疏调适。黍熟获之。桑生，正与黍高平；因以利镰摩地刈之，曝令燥。后有风调，放火烧之，常逆风起火。桑至春生，一亩食三箔蚕。

To plant mulberry trees: In the 5th month (late June), collect (ripened) mulberries, soak them in water, and rub them between the hands. Crush with more water to obtain cleansed seeds. Air to dry. Take 10 亩 *mu* (*mu*≈666.67 m^2) of good land—best to use long-fallowed plots—plough and clean nicely. To every mou, sow a mixture of 3 升 *sheng* (1 *sheng* at that time≈0.17 litre) of mulberry seeds with equal amount of glutinous millet. Both mulberry and millet will germinate about the same time. Hoe them down. Let the mulberry seedlings be properly apart from each other. When the millet ripens, reap the ears first. The mulberry plants will be of the same height as the millet straw, so just cut them both down with a sickle close to the ground. Dry them together in the open. Later, with a good breeze, set fire on the dried plants in a head wind. Next spring, mulberry suckers spring out so that 1 mou will support 3 pan of silkworms.

（《氾胜之书》，石声汉 译）

这一段记述种桑法的文字体现了典籍中量词英译的特点。“亩”和“升”都是中国文化特有的量词,翻译时采用了“直译+汉字+音译+注”的方法,既保留了原语特色,也易于英语读者理解和接受,达到较好的跨文化传播效果。类似的例子还有:

【例20】

作鱼眼沸汤以淋之;令糟上水深一尺许,乃上下水,洽讫。向一食顷,便拔酮取汁煮之。

Heat water until bubbling, pour upon the top of the mash, let water stand about 1 *ch'ih* (23 cm) above the mash residue. Then stir to mix well. After about one meal time, loosen the stoppage of the drains, and ladle out the drained syrup to boil.

(《齐民要术》卷九《饧餔第八十九》,石声汉 译)

除直译、音译和加注外,量词的翻译还可以采用意译的方法,特别是在记述比例的文本中,具体的量词往往根据单位换算译成了比例关系。

【例21】

作鱼酱法:去鳞,净洗,拭令干。如脍法,披破缕切之。去骨。大率:成鱼一斗,用黄衣三升,一升全用,二升作末。白盐二升,黄盐则苦。干姜一升,末之。橘皮一合,缕切之。和令调均,内瓮子中,泥密封,日曝。

For fish-chiang: desquamate, rinse well, wipe to dry, open into fillets, and clear off bones. Use 3 parts (1 pt. full, 2 pts. powdered) yellow mould, 2 parts white salt (brown salt has an unpleasant taste) 1 part powdered dry ginger, and a handful of finely shreded orange-rind, to 10 parts fish fillets. Mix well. Place in a jar, seal tightly with mud, bake in sunshine.

(《齐民要术》卷八《作酱法第七十》,石声汉 译)

原文记述的是作鱼酱的方法。“大率”的意思是比例。量词“合”“升”

“斗”三者的换算方法为:10合=1升,10升=1斗。经过换算,译文用part一词来表示比例关系,“三升”为3 parts,“一斗”为10 parts,更加简洁易懂、利于传播。另外,“合”译为a handful of,也比较形象直观。

【例22】

干蘖末五升,杀米一石。

One part of dried malt-powder will digest twenty parts of cereal.

(《齐民要术》卷九《饧餔第八十九》,石声汉 译)

原文记述的是煮白饧的方法,同样涉及到了比例关系。“斗”和“石”的换算方法为:10斗=1石,所以100升=1石,可算出“五升”和“一石”的比例为1∶20,分别译为one part和twenty parts。

练　习

1.翻译以下典籍原文,特别注意数量词的翻译方法。

1)当令水高下,与重卵相齐。若外水高,则卵死不复出;若外水下卵,则冷气少,不能折其出势。不能折其出势,则不得三七日;不得三七日,虽出“不成”也。“不成”者,谓徒绩成茧、出蛾、生卵,七日不复剖生,至明年方生耳。(《永嘉记》)

2)种名果法:三月上旬,斫取好直枝,如大拇指,长五尺,内著芋头中种之。无芋,大芜菁根亦可。胜种核,核三四年乃如此大耳。(《食经》)

3)预烧落藜、藜、藋及蒿作灰;无者,即草灰亦得。以汤淋取清汁,初汁纯厚太酽,即教花不中用,唯可洗衣。取第三度淋者,以用揉花,和,使好色也。揉花。十许遍,势尽乃止。布袋绞取淳汁,著瓷碗中。取醋石榴两三个,擘取子,捣破,少著粟饭浆水极酸者和之;布绞取渖,以和花汁。若无石榴者,以好醋和饭浆,亦得用。若复无醋者,清饭浆极酸者,亦得空用之。(《齐民要术》)

4)(大豆:)九月中,候近地叶有黄落者,速刈之。(《齐民要术》)

5)(旱稻:)五六月中,霖雨时,拔而栽之。(《齐民要术》)

6)若粪不可得者,五、六月中概种绿豆,至七月、八月,犁掩杀之,如以粪粪田,则良美与粪不殊,又省功力。(《齐民要术》)

7)凡美田之法,绿豆为上;小豆、胡麻次之。悉皆五六月中穙䆉懿反漫掩也种,七月八月,犁稀杀之。为春谷田,则亩收十石;其美与蚕矢熟粪同。(《齐民要术》)

8)熟时,合肉全埋粪地中;直置凡地,则不生,生亦不茂。桃性早实,三岁便结子,故不求栽也。至春既生,移栽实地。若仍处粪中,则实小而味苦矣。(《齐民要术》)

9)七月中,作坑,令受百许束;作麦稈泥泥之,令深五寸,以苫蔽四壁。刈蓝倒竖坑中;下水,以木石镇压令没。热时一宿,冷时再宿,漉去荄,内汁与瓮中。率:十石瓮,著石灰一斗五升,急手抨之,一食顷止,澄清,泻去水。别作小坑,贮蓝淀著坑中;候如强粥,还出瓮中盛之,蓝淀成矣。(《齐民要术》)

10)肉酱法:牛、羊、獐、鹿、兔肉,皆得作。取良杀新肉,去脂细锉。陈肉干者不任用。合脂,令酱腻。晒曲令燥,熟捣绢簁。大率:肉一斗,曲末五升,白盐二升半,黄蒸一升。曝干,熟捣,绢簁。盘上和令均调,内瓮子中。泥封日曝。寒月作之,宜埋之于黍穰积中。(《齐民要术》)

11)余昔有羊二百口,茭豆既少,无以饲。一岁之中,饿死过半;假有在者,疥、瘦、羸、弊,与死不殊;毛复浅短,全无润泽。余初谓家自不宜,又疑岁道疫病。乃饥饿所致,无他故也。……传曰:“三折臂知为良医。”又曰:“亡羊治牢,未为晚也。”世事略皆如此,安可不存意哉?(《齐民要术》)

12)区种瓜:一亩为二十四科。区方圆三尺,深五寸。一科用一石粪;粪与土合和,令相半。以三斗瓦瓮,埋著科中央,令瓮口上与地平。盛水瓮中,

令满。种瓜，瓮四面，各一子。以瓦盖瓮口。水或减，辄增，常令水满。(《氾胜之书》)

13)教民养育六畜，以时种树，务修田畴，滋植桑麻。肥硗高下，各因其宜。丘、陵、坂、险，不生五谷者，以树竹、木。春伐枯槁，夏取果、蓏，秋畜蔬食，冬伐薪、蒸，以为民资。(《淮南子》)

14)常山之东，河汝之间，蚤生而晚杀，五谷之所蕃孰也。四种而五获，中年亩二石，一夫为粟二百石。(《管子》)

2. 将以下两段话译成英文。

1)《齐民要术》大约成书于北魏末年(公元533—544年)，是中国杰出农学家贾思勰所著的一部综合性农学著作，也是中国现存最早最完整的农学名著。全书10卷92篇，系统地总结了六世纪以前黄河中下游地区劳动人民农牧业生产经验、食品的加工与贮藏、野生植物的利用，以及治荒的方法，详细介绍了季节、气候及不同土壤与不同农作物的关系，被誉为“中国古代农业百科全书”。

2)中国古代将农历五月视为“毒月”，直到清代仍是如此。这个月青黄不接，存粮已经吃完，新粮还没有生产出来，人们生计维艰，同时虫毒并作，疫病易行，人们心怀畏惧，禁忌多端。为了生存并生活好，人们以特定习俗活动来表达除毒祈食，趋吉辟邪，如缠朱索(五色丝)、挂长命缕等。在有些地域，父母更将未满周岁的儿童带至外婆家躲藏，以避不吉的毒月。

3. 以下三段分别记述了香泽、手药、香粉的制作方法，根据释文进行翻译。

合香泽法：好清酒以浸香。夏用冷酒，春秋温酒令暖，冬则小热。鸡舌香、藿香、豆蔻、泽兰香，凡四种，以新绵裹而浸之。夏一宿，春秋再宿，冬三宿。用胡麻油两分，猪脂一分，内铜铛中，即以浸香酒和之。煎数沸后，便缓火微煎；然后下所浸香，煎。缓火至暮，水尽沸定，乃熟。以火头内泽中。作声者，水未尽；有烟出无声者，水尽也。

合手药法:取猪脏一具,摘去其脂。合蒿叶,于好酒中痛挼,使汁甚滑。白桃人二七枚,去黄皮,研碎,酒解取其汁。以绵裹丁香、藿香、甘松香、橘核十颗,打碎,著脏汁中。仍浸置勿出,瓷瓶贮之。夜煮细糠汤,净洗面,拭干,以药涂之。令手软滑,冬不皴。

作香粉法:唯多著丁香于粉合中,自然芬馥。亦有捣香末绢筛和粉者,亦有水浸香以香汁溲粉者。皆损色,又费香。不如全著合中也。

(《齐民要术》卷五《种红蓝花及栀子第五十二》)

【释文】

配合香泽的方法:用好的清酒来浸香料。夏天用冷酒;春天秋天,把酒烫暖;冬天要把酒烫到热热的。香料,用鸡舌香、藿香、豆蔻、泽兰香四种;用新丝绵包着,浸在酒里。夏天,过一夜;春天秋天,两夜;冬天,三夜。把两分麻油,一分猪油,放在小铜锅里。加上浸香的酒。煮沸几遍后,将火退小,慢慢地煎。随后再将浸过的香一起煎,更要小火。到晚上酒带来的水煎干了,也不再沸了,就已成功。拿火头淬到香泽里,如果有声音,表示水还没干;出烟而不响,便是水干了。

配合手药的方法:取一副猪胰,把附着的脂肪组织摘掉。加上青蒿叶子,在好酒里面,用力挼揉,让汁液滑腻。用二十七枚白桃仁,剥去黄色的种皮,研碎,用酒浸取汁。用丝绵包裹丁香、藿香、甘松香和十颗打碎了的橘核,一同放在胰子汁里。让它们浸着,不要取出来,搁在瓷瓶里贮藏着。晚上,把煮细糠所得到的汤,将手脸洗净,擦干,将手药涂上。可以使手柔软滑润,冬天不皴裂。

作香粉法:只要在粉盒子里多放些整颗的丁香,自然就芬香馥郁。有人把香捣成末,绢筛筛过和到粉里面的;也有用水浸着香料,用香汁和粉的。都使粉色不白,而且费香。不如整颗放在盒中的好。

延伸阅读

《齐民要术》译文赏析

下文节选自石声汉自译《齐民要术概论》，原文出自《齐民要术》卷一《耕地第一》，记述的是耕地的方法和注意事项。

凡开荒山泽田，皆七月芟艾之。草干，即放火。至春而开垦(根朽省工)。其林木大者，杀之，叶死不扇，便任耕种。三岁后，根枯茎朽，以火烧之(入地尽也)。耕荒毕，以铁齿鳎楱，再遍杷之。漫掷黍穄，劳亦再遍；明年，乃中为谷田。

To reclaim any hilly or swampy waste, first cut down the wild vegetations in the 7th month. When the rubbish has dried a little, set fire to it. Plough to open up next spring—the roots are now rotten, so the ground will be easily workable. If there be any woody growth too big to be hewn away, girdle it to kill. When its leaves give no more canopying effect, the land will be ready for ploughing and sowing. After 3 years' time, set fire to the withered and rotten trunks, then all these wild growth can eventually be purged down beneath. After ploughing, level down twice the clods with an iron-toothed rake. Broadcast glutinous or ordinary panicled millets, harrow twice. The plots will be ready for spiked millet next year.

凡耕，高下田，不问春秋，必须燥湿得所为佳。若水旱不调，宁燥不湿。燥耕虽块，一经得雨，地则粉解。湿耕坚垎，数年不佳。谚曰："湿耕泽锄，不如归去。"言无益而有损。湿耕者，白背速鳎楱之亦无伤；否则大恶也。

When ploughing high or low fields, no matter whether in the spring or autumn, always look to the proper moisture; in years of unsuitable rainfall, it is best to plough when dry but never when wet. Ploughing when dry, the soil comes up in clods which will crumble asunder when moistened by rain. With wet soil, when ploughed up, there will be formed stubborn clods which remain hard for years to come. The proverb that "it is better go home and rest, than to plough the wet and hoe the drenched" means to say that such practices are not only useless, but actually harmful. If the ground has been once ploughed when wet, a good remedy is to draw an iron-toothed rake over it as soon as the surface turns pale. If this is not done, it will definitely be very bad.

春耕,寻手劳;秋耕,待白背劳。(春既多风,若不寻劳,地必虚燥;秋田𡒄实,湿劳令地硬。谚曰:"耕而不劳,不如作暴。"盖言泽难遇,喜天时故也。)

凡秋耕欲深,春夏欲浅;犁欲廉,劳欲再。(犁廉耕细,牛复不疲。再劳地熟,旱亦保泽也。)

Harrow immediately after ploughing in spring; but wait till the surface turns pale in autumn: In spring-time, it is always very windy, ploughed land, being too light, will soon dry out. In autumn, after soaking by successive showers of rain, the soil is already too stickly and hard-beaten; harrowing certainly makes it worse. The proverb says, "If no harrowing goes after ploughing, better to do nothing at all." It means to emphasize that rainfall being scarce, one should utilise it to the utmost advantage.

In autumn, the drill of plough ought to be deep; in spring and summer, shallower. Furrows should be narrow, and harrowing repeated. —Narrow furrows means fine turning, and the draught ox will be less easily tired out. Repeated harrowings would mix the soil better and conserve more moisture even in drought.

秋耕,稀青者为上。(比至冬月,青草复生者,其美与小豆同也。)

初耕欲深,转地欲浅。(耕不深,地不熟;转不浅,动生土也。)

菅茅之地，宜纵牛羊践之(践则根浮)。七月耕之，则死。非七月，复生矣。

In autumn ploughing, the best thing to do is to bury all the green weeds under the furrows. When green grasses grow out in winter time, (turn them down again) the benefit will be the same as if lesser beans had been ploughed down.

Primary ploughing should be deep; later, shallower. Deep primary ploughing mix the soil better, but deep ploughing later may stir up raw subsoil.

Grounds overgrown with rushes should be trodden by cattle and sheep first. Treading turns roots up. Then plough them down in the 7th month, so as to kill them successfully. If not ploughed down in the 7th month, they will revive.

拓展知识

下文是《氾胜之书》的背景介绍，由石声汉撰写。除注释中的繁体字转换为简体字外，其余均为原文摘录。

The Background of *Fan Shêng-Chih Shu*

by Shih Shêng-Han

Pending any disproof, we shall for the time being assume that *Fan Shêng-Chih Shu* is actually a work of Fan Shêng-Chih the agriculturist himself. —that is to say, it is really an ancient agriculturistic book written 2,000 years ago. This, however, does not mean that all the matter recorded in that book is Fan's own personal achievements. Quite on the contrary, what Fan actually did is, we believe, only to register the accomplishments of "the peasants". In other words, we do not take the book *Fan Shêng-Chih Shu* as representing the author's personal inventions or discoveries but rather as faithful reflections of the status of agriculture in the middle Yellow River region at his time. Whatever practice adopted then in such farming, and the principles upon which those practices were based as described in *Fan Shêng-Chih Shu* are certainly the knowledge and experiences of past and contemporary working people. Where Fan's merit resides is to collect, to systematise these experience and knowledge, and to formulate them up to scientific conclusions. This service in itself, nevertheless, is no light work, and surely constitutes a contribution to the human history as a whole.

Therefore, in order to analyse the contents of *Fan Shêng-Chih Shu* we may do better to begin with an attempt to understand the historical as well as the natural background of its author *Fan Shêng-Chih* in respect to agriculture.

From the "Ch'un Ch'iu"[①] age downwards through the "Chan Kuo (Warring States)" period, people were well aware from experience of frequent fightings amongst feudal lords that the production of enough food was of the utmost importance. Wise administration of food, both in production and distribution, became one of the main themata of political philosophy practically in every school of thoughts, and knowledge of agrarian technology hence came to be held in high esteem. The so-called "agriculturistic" school became one so popular and prominent that it finally entered many states and private academies as an important or even "ubiquitous" part. The earliest systematical writing of this "agriculturistic" school now extant is the 6 chapters collectively known as "Shih Jung"[②] (accomplishments of Scholars in principles) in *Lü Shih Ch'un Ch'iu*[③].

The six chapters of *Shih Jung*, one of the six "Luen"[④] (Discussions) in *Lü Shih Ch'un Ch'iu*, constitute a self-consistant series, —probably they are the

---

① 春秋
② 士容
③ 吕氏春秋
④ 论

collective work of the few members who composed a department in Lü's private academy as the "agriculturistic group". The first chapter in *Shih Jung* is a general discussion of what "Jung" (demeanour) of "Shih" (scholars) should be. The second, "*Wu Ta*"[①] (to do the great) relates how, why and what great things should an accomplished Shih (scholar) do. The third, "*Shang Nung*"[②] (farming as the most important), preaches that agriculture is a great affair and therefore best attention should be paid to it, —here for "agriculture" both political administration and technological knowledge are meant. After thus developing in principle the paramount importance of agriculture, the remaining three chapters are dispensed to the technological side of land husbandry. The fourth chapter, "*Jen-Ti*"[③] (capacity of the soil), dealing with productivity of food by land; the fifth, "*Pien T'u*"[④] (working the ground) how to alter the soil to meet farming needs; and finally the sixth "*Shen Shih*"[⑤] (discriminating the season) with the choice of appropriate time for various operations. So, on the whole, we can readily see that there is a consistent trend of thinking behind these 6 chapters. The main thesis is: with good regard to the season and soil productivity, man can win good produce from the ground by his own effort through improvements in farming practices. This is, in my opinion, also the central idea of the ancient agriculturistic school. It is the point-blank outgrowth of materialistic cosmognostics. Later on, the agriculturistic group in the Huai Nan[⑥] academy again laid great emphasis on

---

① 务大

② 上农

③ 任地

④ 辩土

⑤ 审时

⑥ 淮南

agricultural policy and technology with the same underlying idea. And the essays and memorials of the self-styled "*Confucianist*" political scholars Chia I① and Ch'ao Ts'o② (2nd cent.) enlarged upon the same thesis, though using the Confucian phraseology. Fan Shêng-Chih too probably accepted the same trend of thinking. In his book *Fan Shêng-Chih Shu*, especially those sections on the basic principles of farming, the stress on proper choice of time and condition of the soil bears ample evidence of this influence from the earlier ideas.

On the other hand, the "yin yang"③ and "wu hsing"④ hypotheses, though originally representing the naïve cosmological concepts themselves, do somehow smack of mysticism. These concepts, confluent with primitive witchcraft, degenerated in the 3rd century, *viz.* not much later than the time of *Lü Shih Ch'un Ch'iu*, into what is known as "fang shuh"⑤ (magic arts) or necromantic shamanism. Followers of fang shuh, the "fang shih"⑥ (scholars of magic arts) were usually favourite courtiers to the Emperors. The emperor Han Wu Ti, after some unprofitable expansion of his empire westwards (which led to the opening of the "silk road" though) and eastwards, gradually developed a great fancy towards corporal immortality. He was more than once dupe to the fang shih. In the fifty-four years of his reign, especially in the later years, his personal living and his court were so meddled with the fang shih that the government as well as the common people were all impregnated with the corrupted creed of yin-yang and wu-

---

① 贾谊
② 晁错
③ 阴阳
④ 五行
⑤ 方术
⑥ 方士

hsing. The so-called "Confucianistic Scholars", shrewd bureaucratic candidates in reality, easily and deftly adopted this duping craft with the aid of their literary learning. These scholars artfully cooked and garnished the primitive creed with eulogical phrases from classics and formed the art "ts'an wei"① for their own benefit. The succeeding emperors Hsuan Ti②, Yuan Ti③, Ch'eng Ti and Ai Ti④ all followed the steps of their forefather in patronising the ts'an wei scholars. And this tribe of cunning intellectuals thrived more and more to culminate in the reign of Wang Mang⑤ and the Later Han. If Fan Shêng-Chih, himself flourishing during the reign of Ch'eng Ti and occupying then the office of "State Instructor" to be in direct touch with people in the vicinities of the metropolis, should not be armed, or anyhow tinged with such creed of ts'an wei, it is certainly unimaginable and unreasonable. We must remember that it is the *Zeitgeist*. Thus, we can apprehend why and how in *Fang Shêng-Chih Shu*, matters quite materialistic, *viz.* knowledge scientific and technological, as summarised from practical experience of agricultural production, exist side by side with phrases and concepts mystic in nature.

As already said above, for Fan's life and career records are too scanty. It is thus very difficult to form any definite statement on the range and contents of his activities. What we can infer is all in a roundabout way. For instance, we cannot say whether he had ever gone eastward beyond the sharp bent or "elbow" of the

① 谶纬
② 宣帝
③ 元帝
④ 哀帝
⑤ 王莽

Yellow River at Han-ku-kuan[①], —since wayfaring and means of communication at that time were not good, while literature indicates that official persons of the central government were particularly forbidden to travel on private errand within the empire, —probably as a guard against treason. However, we may just leave questions of this sort unsolved and be satisfied with setting up another more important one with certainty: The main spot where Fan's activities in agriculture was once focused was in the Kuan Chung[②] district, as is evidenced by the commentations in *Han Shu* that he taught the public to farm in San Fu.

Natural environment for farming of the Kuan Chung district at that time may by inference be speculatively visualised as follows: (1) For soil, the loess was the most predominant and important. As a soil, the loess has and must therefore have had some shortcomings. Several physical properties in structure and texture were not very favourable for farming, as nitrogen and available phosphorus were rather deficient, and humus content was also low. (2) Rivers and streams were numerous, but many of them were already rather far advanced in senescent decay, and erosions were also getting more and more serious. (3) Forests were gradually receding and being meanwhile periodically devastated. (4) The problem of water-supply for agricultural irrigation was chiefly solved by hydraulic engineering. Canal systems for irrigating purpose were started far back in the 4th century by feudal Ch'in[③] State, and continued with extensions and improvements in the imperial Ch'in and Han. Wells were drilled in vast numbers according to some author; but probably not much used for watering the land. On high lands, it

---

① 函谷关

② 关中

③ 秦

was not easy to drill a well deep enough to reach the very low subterranean water table, and it was also too heavy a work to draw; on low lands, it was easier to build canals than to drill. Besides, the heavy load of soluble salts rendered the wellwater highly undesirable for irrigation. (5) The climatic conditions were also not far different from what they are today. The change in atmospheric temperature was rather great in range, and not infrequently very sudden. Amount of precipitational water was low and very unfavourably distributed, —droughts (severe droughts lasting for two or more years occurred periodically) and floods were both not uncommon; snowfalls borne by northwestern wind was one important item in the total precipitation. (6) As manifested in Fan's book, plants cropped included spiked and glutinous millets, winter and spring barley and wheat, soya and lesser beans, hemp, melon, gourd, taro and even water-darnels. Paddy nice was grown in spots near waterheads. Mulberry tree was widely cultivated, oleiferous crops, besides hemp, included *Perilla* and sesame (the latter introduced rather late?). Maize, cotton and rape were not known yet. Alfalfa and peas were recently brought in from central Asia and not yet cultivated on large scales. Thus, on the whole, the plants sharing in rotation were rather simple. (7) Ox was the chief animal power utilised; horses were less used by the farming people, donkeys and mules were new acquisitions (from the north?), while other animals under domestication included sheep, goats, pigs and fowl. Silkworms were already long kept; —bombyxine excrement thus constituted an important article of fertiliser next only to night soil. Whether stable manure was ever used then, we cannot yet say.

On the whole, agriculture in Kuan Chung at that time is principally well-advanced dry-farming adaptive to an arid climate.

At the same time, one must also well remember that Kuan Chung is the site where once arose the Chou[①] people—the traditionally famous agrarian race in ancient China. Abundant knowledge and experiences on agricultural practice should have surely been handed down, enriched and dispersed among the inhabitants for at least 10 or more centuries. Further, from the period of the Warring States down to the fall of the Ch'in Empire, the Kuan Chung district was always the base to send troops outwards and never a battlefield itself. The good tradition of agricultural technology must thus have been well kept locally. The fall of the Ch'in Empire certainly is a great event in history. But how far that fall disturbed the agrarian people in the countryside is not easy to fathom yet. On the other hand, in the successive reign of the 3rd and 4th emperors of Han (both are supposed to be beloved sovereigns), wounds of wars were said to be well healed. Local agricultural productions had again risen to high levels according to historical records. The 5th emperor Wu Ti had desperately exploited the whole country for food-supply and manpower in his ambitious expansion of the empire. Meanwhile the needs for farm produces also spurred him and his government to do much in improving agriculture. After the consecutive reign of the emperors Chao Ti[②], Hsuan Ti, Yuan Ti, Ch'eng Ti, the smart of tense taxation might have well subdued whilst the fruit of political encouragement in agriculture might have gradually ripened to form a peak of development, especially in a district like the vicinities of the metropolis where experts always assembled.

---

① 周

② 昭帝

练　习

1. According to the text, what is the main focus of the book *Fan Shêng-Chih Shu*?

   A. Mystical concepts and witchcraft.

   B. Ancient political philosophies.

   C. Agricultural technology and practices.

   D. Literary learning and Confucianism.

2. What is the significance of the "Shih Jung" chapters in the context of ancient Chinese agriculture?

   A. They focus on the importance of military strategy in farming.

   B. They provide a systematic approach to agricultural principles.

   C. They emphasize the role of scholars in political administration.

   D. They discuss the influence of yin yang and wu hsing on farming.

3. How does the author describe the influence of the "agriculturistic" school in ancient China?

   A. It was limited to a few private academies.

   B. It was considered a minor part of political philosophy.

   C. It was popular and prominent in many states and academies.

   D. It was primarily focused on mystical concepts and magic arts.

4. What was the main thesis of the ancient agriculturistic school, as discussed in the text?

   A. The importance of political administration in agriculture.

   B. The significance of Confucianist scholars in farming practices.

   C. The influence of yin yang and wu hsing on agricultural policies.

   D. The role of seasonal and soil conditions in agricultural production.

5. How does the text describe the technological side of land husbandry in ancient China?

   A. It discussed principles of soil productivity and working the ground.

   B. It focused on the spiritual aspects of farming practices.

   C. It emphasized the importance of magic arts in agriculture.

   D. It promoted the use of Confucian phraseology in agricultural policies.

6. Based on the information provided in the text, what can be inferred about the role of the "Confucianistic Scholars" in ancient China?

   A. They focused on promoting mystical concepts and witchcraft.

   B. They played a significant role in advancing agricultural policies.

   C. They were primarily concerned with military strategy and warfare.

   D. They were skilled in manipulating primitive cosmological concepts for political gain.

7. According to the text, what was the main animal power utilized in agriculture in the Kuan Chung district?

   A. Horses.

   B. Donkeys.

   C. Oxen.

   D. Mules.

8. What is the significance of the reign of the 3rd emperor of Han in the context of agricultural practices in ancient China?

   A. It led to a decline in agricultural technology in the Kuan Chung district.

   B. It disrupted the transmission of agricultural knowledge and practices.

   C. It resulted in the introduction of new farming techniques from central Asia.

   D. It marked the beginning of a period of high agricultural production levels.

第七章　农业科技典籍句子的翻译

◎本章学习目标◎

1. 了解农业科技典籍句子的特点
2. 理解语序调整、增补、省略的含义和原则
3. 掌握农业科技典籍句子的翻译方法

文言文在句子结构上有其自身特点。比如宾语前置,《岳阳楼记》中的“微斯人,吾谁与归?”后半句可以理解为“吾与谁归”。再比如省略主语、宾语等句子成分。因此,在翻译时要注意调整语序,运用增补和省略等方法。

第一节　语　序

语序调整以达意为目的。《诗经》中的“折柳樊圃”,意思是折下柳条围成园篱,翻译时可保留原文语序,译为 Pick willow twigs to fence off the kitchen garden. 也可调整语序,译为 Fence off the kitchen garden with picked willow twigs. 两句译文都体现了“折柳”和“樊圃”的逻辑关系,但含义略有不同,前者强调的是折柳的目的,后者强调的是樊圃的材料。

一、保留原文语序

【例1】

夫富国多粟,生于农,故先王贵之。

Wealth and food supply of a state are produced by the agricultural industry. So ancient sovereigns attached much importance to agricultural industry.

(《管子·治国第四十八》,翟江月 译)

【例2】

凡耕之本,在于趣时,和土,务粪泽,早锄早获。

The basic principles of farming are: choose the right time, break up the soil, see to its fertility and moisture, hoe early and harvest early.

(《氾胜之书》,石声汉 译)

【例3】

汤有旱灾,伊尹作为区田,教民粪种,负水浇稼。

During the reign of Emperor T'ang, there was a long spell of severe drought. The Prime Minister Yiyin developed the system of "ou-t'ien", i.e. cultivation in shallow pits, taught the people to treat the seeds and to carry water for irrigating the crop.

(《氾胜之书》, 石声汉 译)

【例4】

近来北方多吉贝,而不便纺织者,以北土风气高燥,绵毳断续,不得成缕;纵能作布,亦虚疏不堪用耳。

There is an abundance of raw cotton at the north, but it is inconvenient to spin and weave there; for the high winds and dry atmosphere cause the fibres of cotton to break so frequently that it does not form even threads, yet cloth is made from it, though rather sleazy and uneven and not very serviceable.

（《农政全书》卷三十五《蚕桑广类·木棉》, C. Shaw 译）

【例5】

醋酒为用，无所不入，愈久愈良，亦谓之醯。以有苦味，俗呼苦酒，丹家又加余物，谓为华池左味。

When vinegar and wine are used as drugs, they function well in the body. The longer they are stored, the better the quality. Vinegar is also known as Xi. As it is bitter in taste, it is also called Kujiu (meaning "bitter wine"). Taoist alchemists blend vinegar with other drugs and use it as a seasoning in their processing.

（《本草纲目》，罗希文 译）

二、调整原文语序

【例6】

得时之和，适地之宜，田虽薄恶，收可亩十石。

With the choice of appropriate time and favourable conditions of the soil, a harvest of 10 shih per mou is obtainable even from very poor land.

（《氾胜之书》，石声汉 译）

【例7】

三月榆荚时，有雨，高田可种大豆。

In the 3rd month, when elm-trees are fruiting, sow soya beans on highland fields whenever it rains.

（《氾胜之书》，石声汉 译）

【例8】

茶者，南方之嘉木也。

Tea (botanically termed "camellia sinensis") is a fine plant indigenous to South China.

(《茶经·一之源》,姜怡、姜欣 译)

【例9】

其造具,若方春禁火之时,于野寺山园,丛手而掇,乃蒸,乃舂,乃复以火干之,则又棨、朴、焙、贯、棚、穿、育等七事皆废。

On the following occasions, some tea utensils and apparatus may be omitted. During the Cold Food Festival (at the beginning of spring), people could go to harvest tea leaves in mountains or gardens of some rural temples. With enough hands, the newly picked leaves can be processed straight away for the steaming, pounding, and baking. Thus the following seven tools may be spared:

qi (an awl knife with hard wood handle for stabbing tea-cakes);

pu (a bamboo rope used to string tea-cakes for transportation);

bei (a kind of underground range to bake the tea leaves dry);

guan (a whittled bamboo stick pointing through tea-cakes for baking);

peng (a wooden shelf racked on a bei for tea baking);

chuan (a twine of stringed tea-cakes as a measurement unit);

yu (a wooden structured square case to keep dry stored tea-cakes).

(《茶经·九之略》,姜怡、姜欣 译)

【例10】

棉花密种者有四害:苗长不作蓓蕾,花开不作子,一也。开花结子,雨后郁烝,一时堕落,二也;行根浅近,不能风与旱,三也。结子暗蛀,四也。

There are four disadvantages attending the too close planting of cotton: 1, the branches grow too long and do not blossom, and the flowers do not form seeds; 2, when the seeds are formed, they become mildewed and steamed, and the pods presently drop off; 3, the roots are so near the surface, and so close, that they can not resist the effects of winds and rains, or drought; 4, the seeds are dull looking and wormy.

（《农政全书》卷三十五《蚕桑广类·木棉》, C. Shaw 译）

三、主动句变被动句

【例11】

良田宜种晚，薄田宜种早。良地非独宜晚，早亦无害。薄地宜早，晚必不成实也。

Good lands may be sown late, poor lands must be sown early. Good lands may be sown not only late, there is no harm in doing so early; while late sowing on poor lands gives no crop at all, so they must be sown early.

（《齐民要术》卷一《种谷第三》，石声汉 译）

【例12】

羊羔，腊月正月生者，留以作种；余月生者，剩而卖之。

Lambkins born in the last or the first month of the year should be kept for reproduction; kids of other months may be castrated and sold.

（《齐民要术》卷六《养羊第五十七》，石声汉 译）

【例13】

余闻老农云：棉种必于冬月碾取。谓碾必须晒。……春间生意茁发，不宜大晒也。

I have heard old farmers say "that the cotton seeds for sowing should be rolled out in the winter season, and then dried in the sun; ... and during the spring, when they sprouted, they ought not to be put in the hot sun."

（《农政全书》卷三十五《蚕桑广类·木棉》，C. Shaw 译）

【例14】

炒熟乘热压出油,谓之生油,但可点照;须再煎炼,乃为熟油,始可食,不中点照,亦一异也。如铁自火中出而谓之生铁,亦此义也。

Stir-fry the drug and press it to get oil while it is still hot. This is called crude sesame oil. It can be used for external use or for lighting a lamp. If it is refined again, it becomes prepared oil, which is good for cuisine, but not good for external use or lighting a lamp. Although the stir-fried drug has been used, the product is still called crude sesame oil. This is just like the case when iron is refined, when crude iron is obtained.

(《本草纲目》,罗希文 译)

【例15】

凡采茶,在二月、三月、四月之间。茶之笋者,生烂石沃土,长四五寸,若薇蕨始抽,凌露采焉。茶之芽者,发于丛薄之上,有三枝、四枝、五枝者,选其中枝颖拔者采焉。其日,有雨不采,晴有云不采,晴,采之。蒸之,捣之,拍之,焙之,穿之,封之,茶之干矣。

The second, third and fourth months of the lunar year are a proper time for almost all sorts of tea to get harvested. Tea sprouts shaped like chubby bamboo shoots are usually found on fertile detritus soil. They normally stretch out four to five inches in length, resembling fern sprouts from a vernal land. The best picking time for such tea leaves is before daybreak, when dew is still glittering on them. Thinner leaves in the shape of germinating buds are mostly found in bushes, each twig bearing only three, four or five pieces. Select fleshy ones to pluck. When it comes to the proper weather for harvesting fresh tea, rains are definitely out of the question, and cloudiness should be excluded as well. Only clear and fine days allow for this activity. Following the initial step of plucking, the curing then would go from steaming, pounding, molding, baking, stringing, all the way to packing in a row for fresh tea leaves to be processed ready.

(《茶经·三之造》,姜怡、姜欣 译)

练 习

1.翻译以下典籍原文,特别注意语序的调整。

1)艺麻之如何? 衡从其亩。(《诗经》)

2)中田有庐。疆埸有瓜。(《诗经》)

3)我有旨蓄,亦以御冬。(《诗经》)

4)凡稼泽,夏,以水殄草而芟荑之。泽草所生,种之芒种。(《周礼》)

5)霜降而树谷,冰泮而求获,欲得食,则难矣!(《淮南子》)

6)获,不可不速,常以急疾为务。芒张叶黄,捷获之无疑。(《氾胜之书》)

7)获禾之法:熟过半,断之。(《氾胜之书》)

8)获豆之法:荚黑而茎苍,辄收无疑;其实将落,反失之。故曰:"豆熟于场"。于场获豆,即青荚在上,黑荚在下。(《氾胜之书》)

9)获麻之法:霜下实成,速斫之;其树大者,以锯锯之。(《氾胜之书》)

10)获枲之法:穗勃,勃如灰,拔之。夏至后二十日沤枲,枲和如丝。(《氾胜之书》)

11)茶之为用,味至寒,为饮最宜。精行俭德之人,若热渴、凝闷、脑疼、目涩、四肢烦、百节不舒,聊四五啜,与醍醐、甘露抗衡也。采不时,造不精,杂以卉莽,饮之成疾。(《茶经》)

12)今意创一法:不论冬碾、春碾,收藏、旋买,但临种时,用水浥湿过半刻,淘汰之。其秕者、远年者、火焙者、油者、郁者,皆浮;其坚实不损者,必沉。沉者,可种也。(《农政全书》)

13)总种棉不熟之故,有四病:一、秕,二、密,三、瘠,四、芜。秕者,种不实;密者,苗不孤;瘠者,粪不多;芜者,锄不数。(《农政全书》)

14)棉田,秋耕为良。穫稻后,即用人耕。又不宜耙细,须大垡岸起,令其凝沍。来年冻释,土脉细润。正月初转耕,或用牛转。二月初,再转。此二转,必捞盖令细。(《农政全书》)

2.将以下两段话译成英文。

1)《梦溪笔谈》是笔记体综合著作,作者是北宋沈括。《梦溪笔谈》内容涉及天文、历法、气象、地理、物理、化学、水利、建筑、医药、历史、文学、艺术、军事、法律等诸多领域。《梦溪笔谈》以大篇幅记述自然科学,在自然科学领域的研究成果尤为显著。其科学思想和科学精神也深刻地渗透于物理学、化学、天文学、地理科学、生物学等各个学科之中。北宋的重大科技发明和科技人物,均赖之传世。

2)民以食为天。"食"即粮食,泛指人类生存不可或缺的基本资源或物质条件;"天"比喻最重要的事物或主宰一切的根本因素。古人认为,治国者不仅要知道百姓是君主的"天"、国家的"本",而且要知道百姓的"天"是什么;粮食既是百姓糊口养家、安居乐业不可或缺的基本物质条件,当然也是任何领导集团招抚民众,保障民生不可或缺的基本资源。确保百姓能吃上饭、吃饱饭,确保基本生存资源的供应,是治国安民的一条底线。这是一个非常务实的政治理念。

3.参考释文翻译以下典籍原文。

1)以下一首农事诗选自《诗经》,是秋收后周王祭祀土神和谷神的乐歌,根据释文进行翻译。

畟畟良耜,俶载南亩。播厥百谷,实函斯活。或来瞻女,载筐及筥,其饟伊黍。其笠伊纠,其镈斯赵。以薅荼蓼,荼蓼朽止。黍稷茂止,获之挃挃。积之栗栗,其崇如墉,其比如栉。以开百室,百室盈止,妇子宁止。杀时犉牡,有捄其角。以似以续,续古之人。

(《诗经·周颂·闵予小子之什·良耜》)

【释文】

深耕入土的好犁头，开始耕种向阳田。播种那百类好谷，种子生机满相连。有人前来看望你，载了方筐和圆筥，他的饭是黄小米。他的斗笠真结实，他的犁头真好使。用来除去荼和蓼，荼草蓼草都朽死。小米高粱茂盛长，镰刀收割声吱吱。堆积谷物多又多，它的高像城墙起，排列紧密像梳齿。打开上百储藏库，装满百室好停止，妇子心里才安止。杀那公牛来祭祀，有那弯曲的犄角。延续前人来继续，继续古人讲农事。

2)以下两段分别记述了管理粮价和应对饥荒的措施，根据释文进行翻译。

(1)刘晏掌国计，数百里外物价高下，即日知之。人有得晏一事，予在三司时，尝行之于东南。每岁发运司和籴米于郡县，未知价之高下，须先具价申禀，然后视其贵贱。贵则寡取，贱则取盈。尽得郡县之价，方能契数行下，比至则粟价已增，所以常得贵售。晏法则令多粟通途郡县，以数十岁籴价与所籴粟数高下，各为五等，具籍于主者。今属发运司。粟价才定，更不申禀，即时廪收，但第一价则籴第五数，第五价则籴第一数，第二价则籴第四数，第四价则籴第二数，乃即驰递报发运司。如此，粟贱之地，自籴尽极数；其余节级，各得其宜，已无枉售。发运司仍会诸郡所籴之数计之，若过于多，则损贵与远者；尚少，则增贱与近者。自此粟价未尝失时，各当本处丰俭，即日知价。信皆有术。

(《梦溪笔谈·官政卷十一》)

【释文】

刘晏掌管国家财政时，几百里以外地方的物价涨落，他当天就知道了。有人了解到刘晏的一项措施，我在三司任职时，曾经在东南地区推行过。原来，发运司每年从各州县征购粮食时，事前并不知道粮价的高低，必须让各地开列当地的粮价呈报上来。然后根据各地粮价的高低，价高的就少买，价低的就多买。要在收齐各地的粮价以后，才能核定应购的数字发下执行，往往等公文送到时当地的粮价已经上涨,所以常常高价买粮。刘晏的办法是指令产量多、交通便利的州县,将几十年粮价的高低和收购数量的多少,定为五等,开列清单交给主管机关。现在属于发运司主管。粮价刚一确定,不再呈报,可以立即开仓收购粮食。凡是第一等价格就收购第五等的数量,第五等价格就收购第一等的数量,第二等价格就收购第四等的数量,第四等价格就收购第二等的数量。同时派人把收购情况迅速呈报发运司。这样一来,粮价低的地方,自然收购到最多的粮食;其他各地,也按等级购入适当数量的粮食,于是就避免了不合理的收购。发运司还要把各地已收购到的粮食数量汇总统计,如果收多了,就减少粮价高和路远地方的收购量;如果还不够,就增加粮价低和路近地方的收购量。从此以后,定出的粮价就不会再贻误时机,各自与当地粮食收成好坏相适应,当天就知道粮价。这确实是一项好的措施。

(2)皇祐二年,吴中大饥,殍殣枕路。是时范文正领浙西,发粟及募民存饷,为术甚备。吴人喜竞渡,好为佛事。希文乃纵民竞渡,太守日出宴于湖上,自春至夏,居民空巷出游。又召诸佛寺主首,谕之曰:“饥岁工价至贱,可以大兴土木之役。”于是诸寺工作鼎兴。又新敖仓吏舍,日役千大。监司奏劾杭州不恤荒政,嬉游不节,及公私兴造,伤耗民力。文正乃自条叙所以宴游及兴造,皆欲以发有余之财以惠贫者。贸易饮食、工技服力之人,仰食于公私者,日无虑数万人。荒政之施,莫此为大。是岁,两浙唯杭州晏然,民不流徙,皆文正之惠也。岁饥发司农之粟,募民兴利,近岁遂著为令。既已恤饥,因之以成就民利,此先王之美泽也。

(《梦溪笔谈·官政卷十一》)

【释文】

皇祐二年,吴中一带饥荒严重,饿死的人叠压道路。当时范仲淹主持两浙西路,便发放粮食和劝民间施送饮食慰问灾民,采用的各种方法很周全。

吴地一带的人爱好划船比赛,也喜欢做佛事。范仲淹就鼓励民众开展划船比赛,他每天到湖上摆放宴席,从春天到夏天,老百姓全都离家游玩。他又召集众佛寺的住持,吩咐他们:"灾荒之年工价非常低,可以大力兴建庙宇。"于是许多寺庙的修建工程非常兴旺。又翻修粮库和官员的住处,每天也使用着上千的劳力。监司上奏告发杭州长官荒废政务,嬉游取乐没有节制,以及不管官府还是私人都大兴建造之风,损耗民力。范仲淹于是自拟奏章,逐一申诉大兴宴游和兴造之风的目的,都是想挖掘有余的财力,来救济贫穷的人。从事商业、饮食业、建筑业的人,依赖于公私宴游和土建工程而糊口的人,每天至少有几万人。救荒年措施中没有比这更大的了。这一年,两浙灾区,只有杭州平安无事,老百姓中没有人外出流浪,这都是范仲淹的恩德啊。饥荒之年发放官府的粮食,召集老百姓干有益的事,近些年已列入条令。不仅救济饥荒,又趁机为民间兴利,这真是古圣王的德政啊!

延伸阅读

《天工开物》译文赏析

《天工开物》共十八卷,每卷开头都有一段"宋子曰"作为引言,不仅概述全章内容,也不乏作者的个人表达,既富有科学精神,有充满哲学意蕴。

下文分别为第一卷至第六卷的引言,译本由王义静、王海燕、刘迎春合译。

宋子曰:上古神农氏若存若亡,然味其徽号,两言至今存矣。生人不能久生,而五谷生之。五谷不能自生,而生人生之。土脉历时代而异,种性随水土而分。不然,神农去陶唐粒食已千年矣,耒耜之利,以教天下,岂有隐焉。而纷纷嘉种必待后稷详明,其故何也?

纨袴之子以赭衣视笠蓑,经生之家以农夫为诟詈。晨炊晚馕,知其味而忘其源者众矣。夫先农而系之以神,岂人力之所为哉。

《乃粒第一卷》

Songzi says that we should respect the legendary farmers of antiquity (known as the Divine Agriculturists) who first developed agriculture even though it still remains unknown whether or not those people existed in history. Man cannot survive by himself but must rely on the five grains. The five grains cannot grow out of the ground naturally by themselves, but have to be planted by man. The soil changes as time passes by and as a result the varieties and the nature of crops changes due to the changes of the soil. Otherwise, the first cultivation of edible crops has a history of over one thousand years if we trace it back to the time of the Divine Agriculturalist and the semi-legendary Emperor Tao-tang and the farming techniques would be widely known to all. It was not until the time of Houji that the new varieties of crops are recorded in detail.

The offsprings of aristocrats regarded farmers as convicts, and the scholarly families used the word "farmer" as a curse word. These people have abundant food supply and enjoy the good taste of their food, but are very ignorant of the food sources. Therefore, it is natural and reasonable to regard the agriculture first developed by the legendary farmers as a divine cause.

宋子曰，人为万物之灵，五官百体，赅而存焉。贵者垂衣裳，煌煌山龙，以治天下。贱者裋褐、枲裳，冬以御寒，夏以蔽体，以自别于禽兽。是故其质则造物之所具也。属草木者，为枲、麻、苘、葛；属禽兽与昆虫者，为裘、褐、丝、绵。各载其半，而裳服充焉矣。

天孙机杼，传巧人间。从本质而现花，因绣濯而得锦。乃杼柚遍天下，而得见花机之巧者，能几人哉？“治乱经纶”字义，学者童而习之，而终身不见其形象，岂非缺憾也！先列饲蚕之法，以知丝源之所自。盖人物相丽，贵贱有章，天实为之矣。

《乃服第二卷》

Songzi says that man, whose organs and limbs are the best developed, is the highest of all forms of life on earth. The noble men wear clothes that are decorated with pictures of mountains and dragons and they rule the country. Low-status people wear linen clothes, so as to keep warm in winter and stay covered in summer, which differs them from animals. The raw materials of these clothes are supplied by Nature. Cotton, hemp, Chinese jute, and kudzu vine come from plants, and skin, fur, silk and silk floss come from animals and insects. Each makes up half of the raw materials which are necessary to make clothes.

The skill of weaving is popular throughout the country. People make textiles with flower patterns from raw materials, and even make brocade by dyeing and embroidering. Though spinning and weaving machines have been spread all over the country, how many people have seen the weaving skills of jacquard? Some people get some knowledge about weaving skills from books, but they have never seen the actual methods in their lifetime, which is really a pity. Here, at first, we introduce how to raise silkworms to help readers understand where silk is obtained. It is natural that man and clothes should match; his social status can be marked by his clothes.

宋子曰,霄汉之间云霞异色,阎浮之内花叶殊形。天垂象而圣人则之,以五彩彰施于五色,有虞氏岂无所用心哉?飞禽众而凤则丹,走兽盈而麟则碧。夫林林青衣望阙而拜黄朱也,其义亦犹是矣。君子曰,甘受和,白受采。世间丝、麻、裘、褐皆具素质,而使殊颜异色得以尚焉。谓造物而不劳心者,吾不信也。

《彰施第三卷》

Songzi says that clouds in the sky are colorful, and flowers and leaves on the ground are of various shapes. The sages in ancient times copied the color in nature; they dyed clothes various colors, such as green, yellow, red, white and black with dyes and wore them. Emperor Shun did this on purpose. Among the many birds, the red color of phoenix is the most dazzling; among all the beasts the green of the unicorn—a mythical horned beast is the most special. So many common people wear black clothes; they admire and respect the noble in royal palace who wear yellow and red clothes. Junzi once said, "The bland will absorb a mixture of tastes, and the white will absorb the rainbow colors." All the silk, hemp, fur and woolen stuffs are naturally plain, so they can be dyed different colors which make them valuable. This is the elaborate arrangement of nature.

宋子曰,天生五谷以育民,美在其中,有"黄裳"之意焉。稻以糠为甲,麦以麸为衣。粟、粱、黍、稷,毛羽隐焉。播精而择粹,其道宁终秘也。饮食而知味者,食不厌精。杵臼之利,万民以济,盖取诸《小过》。为此者,岂非人貌而天者哉?

《粹精第四卷》

Songzi says that Nature provides five types of grains to nourish people. Grains are hidden in the yellow chaff, and look as beautiful as if they were in yellow robes. Rice is covered in chaff, wheat is enclosed by bran, and millet and sorghum grains are hidden in feather like husks. It is obvious that people can get fine and polished rice and flour by getting rid of their impurities. For those who are particularly interested in the flavor of food nothing can be too refined. Pestle and mortar are used to grind and polish cereals and are useful to everyone. The desire of refining results in an excessive use of small and humble tools such as the pestle and mortar. How can the inventors be common people? In fact they are geniuses.

宋子曰，天有五气，是生五味。润下作咸，王访箕子而首闻其义焉。口之于味也，辛酸甘苦经年绝一无恙。独食盐禁戒旬日，则缚鸡胜匹，倦怠恹然。岂非天一生水，而此味为生人生气之源哉？四海之中，五服而外，为蔬为谷，皆有寂灭之乡，而斥卤则巧生以待。孰知其所以然？

《作咸第五卷》

Songzi says that there are five elements in nature, water, fire, wood, metal and earth, which are the basic elements of everything. They give birth to five tastes: salty, bitter, sour, spicy and sweet. Among the five elements water is wet; it flows and tastes salty. When King Wu of the Zhou Dynasty visited Qizi, a sage at that time, he learned about the five elements. People can survive even if they do not eat food which tastes either spicy, sour, sweet or bitter for as long as a year. However they cannot do without salt. If people do not eat salt for ten consecutive days, they will become so weak and tired that they can not even tie up a chicken. This proves that water is the creation of Nature, and the salt in water is the life-giving source for human beings. There exist large areas of bare and barren lands where crops and vegetables are not grown, but surprisingly, salt is produced and obtainable everywhere.

宋子曰，气至于芳，色至于艳，味至于甘，人之大欲存焉。芳而烈，艳而艳，甘而甜，则造物有尤异之思矣。世间作甘之味，十八产于草木，而飞虫竭力争衡，采取百花酿成佳味，使草木无全功。孰主张是，而颐养遍于天下哉？

《甘嗜第六卷》

Songzi says that all human beings appreciate fragrance, bright colors, and sweet flavor. Some natural products have a strong fragrance, some are brightly colored, and others taste sweet. These are all creations of nature. Most of the sweet things in the world come from sugar cane, while bees are doing their utmost to gather all sorts of flowers to make honey, thus making sugar cane unable to monopolize the whole contribution of making sugar. What kind of natural force makes sugar canes and honey taste sweet to nourish human beings all over the world after all?

第二节　增　补

由于文言文言简意赅的特点,原文中有其义而无其词的现象较为普遍,翻译时可以根据上下文进行适当的增补,使语义更明确、更完整,使译文更忠实、更通顺。

【例1】

春候地气始通:椓橛木,长尺二寸;埋尺,见其二寸。立春后,土块散,上没橛,陈根可拔。此时。二十日以后,和气去,即土刚。以时耕,一而当四;和气去,耕,四不当一。

In the spring time, to watch for the coming through of the breath of the earth: Sharpen a wooden stake 1 ch'ih and 2 ts'un long, bury 1 ch'ih of it below and let the remaining 2 ts'un appear above the ground level. After lih ch'un, the clods begin to disintegrate, hence will heap up and cover the top of the stake, then old stumps of the previous year could be lightly pulled out. This is the proper time to plough. Twenty days later, the mellow breath of the earth is gone, and the soil hardens. One ploughing in proper time is worth four, but four ploughings will not equal to one after the mellow breath is gone.

(《汜胜之书》,石声汉 译)

原文中的"尺"指"一尺","二寸"指埋入土后剩下的长度,译文中增补了数字1和修饰语remaining。在翻译"土块散,上没橛,陈根可拔"时,增补了三个分句之间的逻辑关系。"此时"意思是到了耕地的合适时间，增补 to plough。最后一句的对比关系通过增补 but 体现出来。

【例 2】

麦生，黄色，伤于太稠。稠者，锄而稀之。

Should young wheat plants look sickly yellow, it is the injury from overcrowding. Hoe to thin them down.

（《氾胜之书》，石声汉 译）

译文中有两处增补：一是增补了条件 if young wheat plants should look…，省略 if，should 置于句首；二是增补了 yellow 的修饰语 sickly，明确说明这种黄颜色是不健康的。

【例 3】

区种，天旱常溉之；一亩常收百斛。

With this system of cultivation in shallow pits, one must always see to watering of his crop in days of drought. A yield of 100 hu per mou may be expected if diligently worked for.

（《氾胜之书》，石声汉 译）

【例 4】

寒月作之，宜埋之于黍穰积中。

If made in winter days, bury the jar in millet chaff to keep warm.

（《齐民要术》卷八《作酱法第七十》，石声汉 译）

【例 5】

冯可宾《岕茶笺》："茶，雨前精神未足，夏后则梗叶太粗。然以细嫩为妙，须当交夏时。时看风日晴和，月露初收，亲自监采入篮。如烈日之下，应防篮内郁蒸，又须伞盖，……"

Extracted from *Guidelines on Luojie Tea* (*Jie Cha Jian*) by Feng Kebin: "Tea is not grown up to its prime before the Grain Rain, and will pass its heyday after the Summer Solstice. To harvest the best, pickers have to wait till the buds bred mellow yet tender in late spring and early summer. When the day is charming with soft breeze, and the moonlit dews retreat, tea lovers could take the advantage and inspect each step to ensure good buds. When the sun is scorching, tea makers should see to it that the baskets are sheltered under a parasol. ..."

(《续茶经·茶之道》,姜怡、姜欣 译)

原文第二、三、四句没有主语,在翻译时分别增补了主语 pickers, tea lovers 和 tea makers,施动者更加明确。类似的例子还有:

【例6】

稀不如密者,就极瘠下田言之,所谓瘠田欲稠也。田之肥瘠,在粪多寡,在人勤惰耳。已则瘠之而稠之,自令薄收,非最下惰农,当作此语耶?

To say that to plant them apart is not so good as close, is to impoverish the fields very much; and he who wishes to have lean fields may plant them close. Rich or lean soil depends upon the quantity of manure, and the industry of the workmen; and he will have poor fields who sows close, and get himself a lean ingathering. Was it an industrious husbandmen that spake in this way?

(《农政全书》卷三十五《蚕桑广类·木棉》, C. Shaw 译)

【例7】

俗传胡麻须夫妇同种则茂盛。故本事诗云:胡麻好种无人种,正是归时又不归。

Legend says Huma should be planted by both husband and wife if it is to prosper. Therefore, in the book *Benshi Shi*, there was a poem saying: With good seed of Huma, the wife could not plant it, As the one (the husband) is not yet back home.

(《本草纲目》,罗希文 译)

【例8】

(稻)此得天地之和,高下之宜,故能至完。伐取得时,故能至坚也。

Rice absorbs the harmonic Qi from the heavens and the earth and grows in the desirable places in terms of the height of land. That is why it can be perfect. Besides, it is gathered in in the right time, so its straw is hard in texture.

(《黄帝内经·素问》,李照国 译)

练　习

1. 翻译以下典籍原文,特别注意增补的使用。

1)草木未落,斧斤不入山林。(《淮南子》)

2)取麦种:候熟,可获,择穗大强者,斩,束,立场中之高燥处,曝使极燥。无令有白鱼,有,辄扬治之。取干艾杂藏之,麦一石,艾一把;藏以瓦器竹器。顺时种之,则收常倍。(《氾胜之书》)

3)取禾种:择高大者,斩一节下,把,悬高燥处,苗则不败。(《氾胜之书》)

4)种,伤湿郁热,则生虫也。(《氾胜之书》)

5)牵马,令就谷堆食数口,以马践过。为种,无虸蚄等虫也。(《氾胜之书》)

6)欲知岁所宜:以布囊盛粟等诸物种,平量之,埋阴地。冬至后五十日,发取量之。息最多者,岁所宜也。(《氾胜之书》)

7)种,欲截雨脚;若不缘湿,融而不生。(《齐民要术》)

8)麻欲得良田,不用故墟。故墟亦良,有䕸叶夭折之患。(《齐民要术》)

9)食瓜时,美者收取。即以细糠拌之。日曝,向燥,挼而簸之,净且速也。(《齐民要术》)

10)凡田,来年拟种稻者,可种麦。拟棉者,勿种也。谚曰:歇田当一熟。言息地力,即古代田之义。若人稠地狭,万不得已,可种大麦或稞麦,仍以粪壅力补之,决不可种小麦。凡高仰田,可棉可稻者,种棉二年,翻稻一年,即草根溃烂,土气肥厚,虫螟不生。多不得过三年,过则生虫,三年而无力种稻者,收棉后,周田作岸,积水过冬;入春冻解,放水候乾。耕锄如法,可种棉,虫亦不生。(《农政全书》)

2.将以下两段话译成英文。

1)《礼记》作为一部以儒家礼论为主的论文汇编,其内容主要是记载和论述先秦的礼制,记录孔子和弟子等的问答,涉及政治、法律、道德、哲学、历史、祭祀、文艺、日常生活、历法、地理诸多方面,集中体现了先秦儒家的政治、哲学和伦理思想,小至修身、齐家,大至治国、平天下,尽显我国礼仪之邦的风范。

2)仓廪实则知礼节,衣食足则知荣辱。"仓廪"是古代储藏米谷的地方或设施。"仓廪实""衣食足"指粮食储备充足,民众不愁吃穿,代指人们生产、生活所需的物质条件非常充足,即物质文明发展到一定阶段;"礼节""荣辱"指社会的礼仪规矩和内心的道德准则,包括了制度文明和精神文明。这句话揭示了物质文明和制度文明、精神文明之间的关系:物质文明是制度文明和精神文明产生的基础和条件,制度文明和精神文明是物质文明发展到一定阶段的产物。如果民众的基本生活条件都得不到保障,即使有良好的制度也难为人们所遵循,人们的精神品格也不可能得到提升。在任何时候,物质文明建设都应当成为治国理政的基本要务。这是一种非常务实的治国理念。

3.参考释文翻译以下典籍原文。

1)下文记述了礼俗起源和饮食之间的联系,根据释文进行翻译。

夫礼之初,始诸饮食。其燔黍捭豚,污尊而抔饮,蕢桴而土鼓。犹若可以致其敬于鬼神。及其死也,升屋而号,告曰:"皋!某复!"然后饭腥而苴孰。故天望而地藏也,体魄则降,知气在上。故死者北首,生者南乡。皆从其初。

昔者先王，未有宫室，冬则居营窟，夏则居橧巢。未有火化，食草木之实、鸟兽之肉，饮其血，茹其毛。未有麻丝，衣其羽皮。后圣有作，然后脩火之利，范金合土，以为台榭、宫室、牖户。以炮以燔，以亨以炙，以为醴酪；治其麻丝，以为布帛，以养生送死，以事鬼神上帝，皆从其朔。

故玄酒在室，醴盏在户，粢醍在堂，澄酒在下。陈其牺牲，备其鼎俎，列其琴瑟管磬钟鼓，修其祝嘏，以降上神与其先祖。以正君臣，以笃父子，以睦兄弟，以齐上下，夫妇有所。是谓承天之祜。

（《礼记·礼运》）

【释文】

礼，最初产生于饮食行为。先民把黍米放在烧石上烧熟，把猪肉撕开放在烧石上烧熟，在地上掘坑盛水当作酒樽，用双手捧着当酒杯来喝，用泥土做成鼓槌，敲打泥土做的鼓当作鼓乐。简陋如此，仍然能向鬼神致以敬意。到他们死的时候，活着的人登上屋顶对着天空大声喊叫，他们喊道："噢！某人回来呀！"死者却不能复生。他们就把生米放入死者口中，再为死者包些熟肉，然后下葬。之所以招魂时望着天，身体却埋在地下，是因为身体沉重下降，可知气轻而上浮。身体埋入地下，灵魂却在天上。北方是阴，所以死者埋时头向北；南方是阳，所以活人以南为尊。这都是在遵从最初的习惯。

从前，先代君王没有宫殿房屋，冬天就住在用土营造的窟穴里，夏天就住在用柴草搭成的窝巢里。那时，还不知用火，只能生吃草木的果实和鸟兽的肉，喝鸟兽的血，连毛也吃下去。也没有麻和丝，穿的是鸟羽和兽皮。后世有圣人出来，然后才知道利用火的热力，用模子浇铸金属，调和泥土烧制陶器、砖、瓦，用来建造台榭、宫室、门窗。用火来烧烤烹制熟食，酿造醴酒、乳酪；处理麻丝，用它们织成麻布和丝绸。用这些东西来供养生者，安葬死者，祭祀鬼神上帝。后世的人们在这些方面都是在遵从原始的做法。

因为遵从原始,所以祭祀时把最古的玄酒放在地位最尊的屋内,醴、盏放在户内,粢醍放在行礼的堂上,清酒则放在堂下。摆好牺牲备齐鼎俎,安排琴瑟管磬钟鼓,撰写主人告神的祝辞,尸致福于主人的嘏辞,用来迎接上神和先祖的降临。在祭祀的进程中,使君臣大义得以辨正,父子亲情得以加深,兄弟得以和睦,上下得以沟通,夫妇各有地位。这就叫作承奉上天的福佑。

2)以下三段讲述了三个关于茶的小故事,根据释文进行翻译。

《神异记》:"余姚人虞洪,入山采茗,遇一道士,牵三青牛,引洪至瀑布山,曰:'予,丹丘子也。闻子善具饮,常思见惠。山中有大茗,可以相给,祈子他日有瓯牺之余,乞相遗也。'因立奠祀,后常令家人入山,获大茗焉。"

《异苑》:"剡县陈务妻,少与二子寡居,好饮茶茗。以宅中有古冢,每饮辄先祀之。二子患之曰:'古冢何知,徒以劳意。'欲掘去之,母苦禁而止。其夜梦一人云:'吾止此冢三百余年,卿二子恒欲见毁,赖相保护,又享吾佳茗,虽潜壤朽骨,岂忘翳桑之报。'及晓,于庭中获钱十万,似久埋者,但贯新耳。母告二子,惭之,从是祷馈愈甚。"

《广陵耆老传》:"晋元帝时有老姥,每旦独提一器茗,往市鬻之,市人竞买,自旦至夕,其器不减。所得钱散路旁孤贫乞人,人或异之。州法曹絷之狱中,至夜,老姥执所鬻茗器,从狱牖中飞出。"

《茶经·七之事》

【释文】

据《神异记》中描写:"一次,有位叫虞洪的余姚人到山里采茶,遇见一个道士牵着三条青牛。他引着虞洪到了瀑布山,说:'我就是丹丘子。早就听说你善于烹茶,真希望我也能有幸得以品尝。这山中有大茶树,茶叶任你采摘,只望你日后有多余的茶汤能送些给我。'回家后,虞洪便用茶祭奉丹丘子。之后他常令家人进山寻访,果然找到了丹丘子所说的大茶树。"

《异苑》中有这样的记载:"剡县人陈务的妻子年轻守寡,带着两个儿子过活。陈妻喜欢饮茶,她家的院子里有个古坟,每次饮茶前她都先用茶祭祀一下古坟。两个儿子对此很不以为然,说:'古坟知道什么啊,你这不是白费心思吗?'他们甚至想把古坟掘掉,幸得母亲苦苦相劝,才没得手。就在这天夜里,陈妻梦见一个人对她说:'我住在你院子这座坟里已有三百余年了,你的两个儿子常想毁掉它,全承蒙你的保护,你还给我美茶享用,我虽是地下朽骨,也知晓报恩。'第二天天亮,陈妻果然发现在庭院中放有十万铜钱,看样子铜钱像是埋在地下很久了,但穿钱的绳子却是新的。陈妻将此事告诉两个儿子,他们心里很感愧疚。从此以后,母子常常一起祭祷古坟,仪式进行得愈加庄重。"

据《广陵耆老传》记载:"晋元帝时,有位老婆婆,她每天清晨都独自提着盛茶的器具到市场上卖茶,人们争相购饮。可奇怪的是,从早到晚,她壶里的茶却丝毫不见减少。老婆婆把卖茶所得的钱全部分给路旁的孤儿与乞讨的穷人。有人觉得这事蹊跷,把她通报给州府。于是,州执法官把她抓了起来,囚入狱中。深夜,老人竟带着自己卖茶的器皿,越过监狱的铁窗飞走了。"

延伸阅读

《天工开物》译文赏析

下文分别为第七卷至第十二卷的引言,译本由王义静、王海燕、刘迎春合译。

宋子曰,水火既济而土合。万室之国,日勤一人[①]而不足,民用亦繁矣哉。上栋下室以避风雨,而瓴建焉。王公设险以守其国,而城垣、雉堞,寇来

① 此处应为"千人"。典出《孟子·告子下》:"万室之国一人陶,则可乎?曰:不可,器不足用也。"在明朝,陶瓷业迅猛发展,宋应星借用典故,将"一人"改为"千人",运用夸张手法形容当时兴旺的陶瓷业。

不可上矣。泥瓮坚而醴酒欲清,瓦登洁而醯醢以荐。商周之际,俎豆以木为之,毋以质重之思耶。后世方土效灵,人工表异,陶成雅器,有素肌、玉骨之象焉。掩映几筵,文明可掬。岂终固哉!

《陶埏第七卷》

Songzi says that through the interaction of water and fire, clay can be burnt into ceramics. According to the ancients, within an area of ten thousand households, pottery made by one person is not enough for people's need. It is obvious that pottery is widely used among the folk. Roofs should be covered with tiles to make them weatherproof. Monarchs set dangers and obstacles to protect their countries. Walls and parapets should be made with bricks to keep enemy out. Strong earthen urns can keep the wine fragrant. Clean goblets are suitable to hold offerings for sacrifice. Between the Shang and Zhou dynasties, sacrificial dishes were made of wood. It is not because people love them, but because they lack the related skills. Later on, people of different places rushed to come up with better techniques. With the techniques changing with each passing day, wooden ware was replaced with polished pottery. Some of the pottery is as thin as paper, some is as white as white jade. These wares sparkle in quiet retreats or at festive boards, a concrete sign of civilized life. So from this, how can we say that things are the same all the time?

宋子曰,首山之采,肇自轩辕,源流远矣哉。九牧贡金,用襄禹鼎。从此火金功用日异而月新矣。夫金之生也,以土为母。及其成形而效用于世也,母模子肖,亦犹是焉。精粗巨细之间,但见钝者司舂,利者司垦,薄其身以媒合水火而百姓繁。虚其腹以振荡空灵而八音起,愿者肖仙梵之身,而尘凡有至象。巧者夺上清之魄,而海寓遍流泉。即屈指唱筹,岂能悉数,要之人力不至于此。

《冶铸第八卷》

Songzi says that ever since the time of the Yellow Emperor, copper mining and casting of three-legged tripod or four-legged cauldron for cooking have been conducted in Shoushan. In China, mining has had a very long history. In the time of Emperor Yu of the Xia Dynasty, metals were presented to the Emperor for his Great Tripod by the magistrates of the nine provinces. Ever since then, the craftsmanship of metal casting by fire improved with time. Metal is produced from earth when it is in its natural state. When it is made into implements for people, it is produced like earthen molds. The implements vary in quality and size. People use these implements widely: from one that is blunt as a mortar for pounding grains to one that is as sharp as a ploughshare for plowing to one as thin as an iron wok for containing water and cooking. A big bell is made from metal and made into a hollow shape which creates harmonious sounds that fill the air. Chiliasts create figures with metal so that the images of Buddha can be seen in this mortal world. The surface of an exquisite copper mirror can be glazed so that it will be even brighter than sunlight and moonlight. Metal coins are circulated throughout the country. The uses are so numerous that it is impossible to count them. On all accounts, human efforts can do more than these.

宋子曰,人群分而物异产,来往贸迁以成宇宙。若各居而老死,何藉有群类哉?人有贵而必出,行畏周行。物有贱而必须,坐穷负贩。四海之内,南资舟而北资车。梯航万国,能使帝京元气充然。何其始造舟车者,不食尸祝之报也? 浮海长年,视万顷波如平地,此与列子所谓御泠风者无异。传所称奚仲之流,倘所谓神人者非耶?

《舟车第九卷》

Songzi says that the society is formed and maintained by the mobility of different groups of people, goods and services from different places. Indeed, if people cooped themselves in their own homes, without any new contacts, how could it be possible for the society to come into being? Decent people would always like to travel around, but they might be set back by the long, tiring journey; some cheap goods are necessary for everyday life, but are scarce in the local area and have to be transported from other places. Almost anything from the journey to the transport can hardly be achieved without carts, boats and other means of transport. In our country, people in the South rely heavily on boats and in the North on carts. The capital city enjoys great prosperity from travel and trade between different parts of the country. Why shouldn't the first life who started building boats and carts receive their due respect? In my opinion, the boatman, who has spent many years of his life traveling on the sea and can now take ease on the bumpy waves as if he were on the flat land, bears no difference from the great Taoist Liezi who was said to have the capacity to fly with the wind. By the same token, it is perfectly right to regard people like Xi Zhong—who made the first cart—as great saints.

宋子曰,金木受攻而物象曲成。世无利器,即般、倕安所施其巧哉? 五兵之内、六乐之中,微钳锤之奏功也,生杀之机泯然矣。同出洪炉烈火,大小殊形。重千钧者系巨舰于狂渊;轻一羽者透绣纹于章服。使冶钟铸鼎之巧,束手而让神功焉。莫邪、干将,双龙飞跃,毋其说亦有徵焉者乎?

《锤锻第十卷》

Songzi says that metals and timber are made into various appliances through machining. Without right-handed tools, artisans even like Luban and Chui, in ancient China, could not demonstrate their wonderful skills. Without clamps or hammers, the processes of producing various weapons and metallic musical instruments could not be finished. All tools and appliances are forged by blazes in smelters, but the tools and appliances vary in shapes and sizes. Anchors as heavy as three thousand jin anchor in raging waves; needles as light as a feather make floral patterns on a ceremonial robe. These wonderful arts and crafts outshine those of casting bells and tripods. Yet if we assign all the credit of superb smelting and casting to supernatural forces, it seems that proof of this can be found in the story of Moye and Ganjiang swords, and in the rise heavenward of the two famous swords which seemed to turn into two dragons after being wielded. This story may be well-founded.

宋子曰,五行之内,土为万物之母。子之贵者岂唯五金哉!金与火相守而流,功用谓莫尚焉矣。石得燔而咸功,盖愈出而愈奇焉。水浸淫而败物,有隙必攻,所谓不遗丝发者。调和一物以为外拒,漂海则冲洋澜,粘甃则固城雉。不烦历候远涉,而至宝得焉。燔石之功,殆莫与之京矣。至于矾现五色之形,硫为群石之将,皆变化于烈火。巧极丹铅炉火。方士纵焦劳唇舌,何尝肖像天工之万一哉!

《燔石第十一卷》

Songzi says that among the five elements, gold, wood, water, fire and earth, earth is the origin of all the living things on earth. Valuables obtained from earth are more than just metals. When metals and fire interact with each other and the metals melt, thus utensils can be made. Its use cannot be compared and surpassed. However, calcining non-metal ores can also have the same function. If water penetrates the hull of a ship, it will harm the ship. To make matters worse, water can enter any tiny crack, even as tiny as a hair. However, filling the cracks with lime can prevent water from coming in. The ship can battle waves and travel overseas. Walls built by bricks which are made from lime are very strong. The material can be got easily nearby. Therefore, calcining stones are very useful.

Furthermore, the alum of five colors and the masterly qualities of sulphur result from the application of intense heat. Such skills climax in the distilling of litharge. However, even though alchemists boast laboriously, how can their ability match even one-thousandth of the power of nature?

宋子曰,天道平分昼夜,而人工继晷以襄事,岂好劳而恶逸哉?使织女燃薪、书生映雪,所济成何事也?草木之实,其中蕴藏膏液,而不能自流。假媒水火,凭借木石,而后倾注而出焉。此人巧聪明,不知于何禀度也。人间负重致远,恃有舟车。乃车得一铢而辖转,舟得一石而罅完,非此物之功也不可行矣。至范蔬之登釜也,莫或膏之,犹啼儿之失乳焉。斯其功用一端而已哉。

《膏液第十二卷》

Songzi says that Nature divides time into day and night. However, people work day and night by using oil lamps. Can we conclude by this phenomenon that people like working and hate leisure? If women were to weave by light of the firewood and students were to study by the glow of snow, would they be successful? The seeds of grasses and trees are rich in oil, which can't flow out by itself. When people do something to the seeds of grasses and trees with the help of the forces of water and fire and the pressure of wooden and stone utensils, the oil will come out. We don't know how the skills and the wisdom of the ancient people are handed down from one generation to another. People transport the heavy objects to distant places by using boats and carts. It takes a little oil for the wheels to move. It takes a lot of oil to jam the chink of a boat. Thus, neither carts nor boats can move without oil. Cooking without oil is like letting an infant go without milk. This is only one of the functions of oil.

第三节 省 略

文言文特有的语气助词，如夫、盖、也、焉、乎、矣、哉等，在翻译时往往省略。原文中重复的语义，或者上下文中非常明确的语义，为避免冗余，在翻译时也可以省略。

【例1】

夫日回而月周，时不与人游。故圣人不贵尺璧，而重寸阴。时难得而易失也。

The sun and moon move around on their courses, time waits for nobody. Therefore the sages value a whole foot of jade less than one inch of sun-shadow. Time is easily lost and hard to get.

（《淮南子》，石声汉 译）

句首的"夫"表示发表议论，句尾的"也"表示肯定陈述，无实际意义，翻译时均省略。类似的例子还有：

【例2】

行其田野，视其耕芸，计其农事，而饥饱之国可以知也。

Travel through the fields of a state to inspect how they are plowed and weeded and to calculate the output of grain, and then whether people of the state have enough food or are suffering from hunger will be clear.

（《管子·八观第十三》，翟江月 译）

【例3】

凡麦田，常以五月耕。六月，再耕。七月，勿耕！谨摩平以待种时。五月耕，一当三；六月耕，一当再；若七月耕，五不当一。

Fields intended for wheat should always be ploughed in the 5th month.

Plough again in the 6th month. Don't plough in the 7th month, but diligently harrow it level, and wait for sowing. One ploughing in the 5th month is worth three; one in the 6th month, two; but five in the 7th is not worth one.

(《氾胜之书》,石声汉 译)

此例中划线部分的"耕"出现了三次,译文中只在第一个分句使用了ploughing,后面均用数字直接指代,符合英语表达习惯。第二个分句中的is worth也省略,避免重复。类似的例子还有:

【例4】

上田,夫食九人。下田,夫食五人。可以益,不可以损。一人治之,十人食之,六畜皆在其中矣。此大任地之道也。

A farmer working on the best land should be able to provide food for nine persons, and a farmer working on the poorest land, for five persons. The amount of food so provided should increase and not decrease. This formula of one person working on the farm to feed ten persons includes the feeding of animals. This is the way of making the best use of land.

(《吕氏春秋·士容论第六·上农》,汤博文 译)

此例除了省略should be able to provide food之外,还运用了增补。"可以益,不可以损"增补了主语the amount of food so provided,"一人治之,十人食之"增补了同位成分this formula of。

【例5】

锄棉者,功须极细密。昔有人佣力锄者,密埋钱于苗根。锄者贪觅钱,深细爬梳,棉则大熟。

When you hoe cotton, do it carefully. In former times the master used to secrete cash about the roots of the plants, and then tell the laborers to find them, by which device the hoeing was well done, the soil being sifted and "combed out," and the cotton consequently most abundant.

（《农政全书》卷三十五《蚕桑广类·木棉》，C. Shaw 译）

练　习

1. 翻译以下典籍原文，特别注意省略的使用。

1）春伐枯槁，夏取果、蓏，秋畜蔬食，冬伐薪、蒸，以为民资。（《淮南子》）

2）耕之为事也劳，织之为事也扰，扰劳之事而民不舍者，知其可以衣食也。（《淮南子》）

3）黄白土宜禾；黑坟，宜黍、麦；赤土，宜菽也；污泉宜稻。（《孝经援神契》）

4）饮真茶令人少眠。（《广志》）

5）今人种麦杂棉者，多苦迟，亦有一法：预于旧冬耕熟地，穴种麦。来春，就于麦陇中，穴种绵。但能穴种麦，即漫种棉，亦可刈麦。（《农政全书》）

6）行其山泽，观其桑麻，计其六畜之产，而贫富之国可知也。（《管子》）

2. 将以下两段话译成英文。

1）《淮南子》又名《淮南鸿烈》，是西汉淮南王刘安（前179—前122）及其门客集体撰写的一部哲学著作。内容原分为内中外篇，现仅存内篇21篇。书中以道家思想为主，糅合了儒法阴阳五行等家的思想，所以又一般认为它是杂家著作。书中保存了不少自然科学史料和神话寓言故事，也记载了不少秦汉间的轶事，内容比较丰富。

2）"天地之气"是天地所蕴含的气。"气"是构成一切有形之物的基本材料，充盈流动于天地之间。"天地之气"没有具体的形状，但有阴阳两种相反的属性。"天之气"属阳，"地之气"属阴。"天地之气"的运行遵循着阴阳变化的法则。"天地之气"的交互作用推动着昼夜寒暑、风雨雷电的天象变化，也决定了万物的生成与变化。

3.参考释文翻译以下典籍原文。

1)下文记述了中国古代的进餐礼仪,根据释文进行翻译。

凡进食之礼,左殽右胾,食居人之左,羹居人之右。脍炙处外,醯酱处内,葱渫处末,酒浆处右。以脯脩置者,左朐右末。

客若降等,执食,兴,辞,主人兴,辞于客,然后客坐。主人延客祭:祭食,祭所先进。殽之序,遍祭之。三饭,主人延客食胾,然后辩殽。主人未辩,客不虚口。

侍食于长者,主人亲馈,则拜而食;主人不亲馈,则不拜而食。共食不饱,共饭不泽手。

侍饮于长者,酒进则起,拜受于尊所。长者辞,少者反席而饮。长者举未釂,少者不敢饮。

《礼记·曲礼》

【释文】

向客人进餐的礼仪是,把带骨头的熟肉块放在左边,把熟的纯肉块放在右边,饭食放在客人左边,羹汤放在客人右边。切成薄片的肉和烤肉放在外侧,醋和酱放在内侧,蒸葱等拌料放在末端,酒浆等饮品放在右边。如果另加脯、脩等干肉制品,中间弯曲的部分应放在左,末端放在右。

如果客人地位低于主人,就应拿着饭站起来表示辞让,主人也要起身,向客人推辞,然后请客人就座。主人引导客人进行餐前祭礼。行餐前祭礼的时候,祭祀的食物要从先端上的食物开始,依次遍祭各种食物。客人吃过三口饭之后,主人要引导客人吃纯肉块,然后吃带骨的肉块。主人还没吃完,客人不要饮酒漱口。

陪长者吃饭,主人亲自取菜肴给你时,应行拜礼而后再吃;主人不亲自为你进食,那不必拜就可以开始吃。与别人共用餐具一起吃饭,不可只顾自己吃饱,要提前把手洗干净不要临吃的时候揉搓手。

陪侍长辈饮酒，如果长辈递酒给晚辈，晚辈应起身到放酒樽的地方拜谢长辈而后接受。长辈表示谦虚和推辞，而后晚辈返回座席喝酒。长辈没有举杯喝干之前，晚辈不可以先喝。

2)下文记述了夏服布料的制作过程，根据释文进行翻译。

凡苎麻无土不生。其种植有撒子、分头两法。(池郡每岁以草粪压头，其根随土而高，广南青麻撒子种田茂甚。)色有青、黄两样。每岁有两刈者、有三刈者，绩为当暑衣裳、帷帐。凡苎皮剥取后，喜日燥干，见水即烂。破析时则以水浸之，然只耐二十刻，久而不析亦烂。苎质本淡黄，漂工化成至白色。(先取稻灰、石灰水煮过，入长流水再漂，再晒，以成至白。)纺苎纱，能者用脚车，一女工并敌三工。唯破析时穷日之力只得三五铢重。织苎机具与织棉者同。凡布衣缝线、革履串绳，其质必用苎纠合。

凡葛蔓生，质长于苎数尺。破析至细者，成布贵重。又有苘麻一种，成布甚粗，最粗者以充丧服。即苎布有极粗者，漆家以盛布灰，大内以充火炬。又有蕉纱，乃闽中取芭蕉皮析、绩为之，轻细之甚，值贱而质枵，不可为衣也。

(《天工开物·乃服第二卷》)

【释文】

苎麻到处都可以生长。其种植有播种和分根两种方法。(池州每年将草粪压在根部，麻根顺着压土而长高。广南的青麻以种子撒在田地，长得颇茂盛。)苎麻有青、黄两种颜色。每年有收割两次、三次的，织成夏天用的衣服和帷帐。苎麻剥皮后，最好在阳光下晒干，否则见水就烂。将麻皮撕破时要用水浸泡，但只能浸二十刻(五小时)，浸久时不撕也要烂。苎麻本是淡黄色的，经过漂洗才成为白色。(先用稻草灰水或石灰水煮过，再在流动的水中漂洗，晒干后就成白色。)纺苎纱的能手用脚踏纺车，一女工可抵三人。但撕裂麻皮则一日只得三五铢重纤维。织苎麻的机具与织棉相同。缝布衣的线和作皮鞋的串绳，都用苎麻搓成。

葛是蔓生的，其纤维比苎麻长数尺。用破析得很细的葛纤维织布，十分贵重。还有一种苘麻，织成布较粗，最粗的布用作丧服。即使是苎布，也有很粗的，漆工用以蘸灰擦磨漆器，而宫内则用以作火把。还有一种蕉纱，是福建地区取芭蕉的韧皮破析、纺织而成，轻细之甚，不值钱也不结实，不堪做衣服。

延伸阅读

《天工开物》译文赏析

下文分别为第十三卷至第十八卷的引言，译本由王义静、王海燕、刘迎春合译。

宋子曰，物象精华、乾坤微妙，古传今而华达夷，使后起含生目授而心识之，承载者以何物哉？君与臣通，师将弟命，凭借咕咕口语，其与几何？持寸符、握半卷，终事诠旨，风行而冰释焉。覆载之间之借有楮先生也，圣顽咸嘉赖之矣。身为竹骨与木皮，杀其青而白乃见，万卷百家，基从此起，其精在此，而其粗效于障风、护物之间。事已开于上古，而使汉晋时人擅名记者，何其陋哉。

《杀青第十三卷》

Songzi says that the descendants are able to learn and understand the essence of the world and incredible wonders of nature through reading, which has been passed down from ancient times till now, from the Central Plains to the border areas. However, what materials were used for the recording? How much information could be delivered between lord and liegeman, or teacher and student if we just depended on the spoken languages? But by using a piece of paper or half a volume of writing, teaching can be achieved and government orders can be carried out very easily. There is a kind of paper known as Mr. Zhu's paper across the country. It is beneficial to everyone no matter whether he is smart or not. Paper is made from bamboo sticks and cortices whose green barks are removed to make white paper. Thousands of volumes of books of specialists and schools of thought are handed down by paper. The refined paper is used for this purpose, while the rough is for window stuffing and wrapping. Paper making originated in ancient times, while some believe it was invented by some individuals in the Han or Jin Dynasty. What a naive opinion it is!

宋子曰,人有十等,自王公至于舆台,缺一焉而人纪不立矣。大地生五金以利天下与后世,其义亦犹是也。贵者千里一生,促亦五六百里而生。贱者舟车稍艰之国,其土必广生焉。黄金美者,其值去黑铁一万六千倍,然使釜鬵、斤斧不呈效于日用之间,即得黄金,值高而无民耳。贸迁有无,货居《周官》泉府,万物司命系焉。其分别美恶而指点重轻,孰开其先,而使相须于不朽焉?

《五金第十四卷》

Songzi says that people are classified into ten classes ranging from noble kings and dukes to sedan chair bearers, without any of whom the hierarchy will not be established. In a similar manner, the earth produces different metals for the benefits of all the human beings. The precious metals are found once in a thousand li or at least five hundred li. The less valuable metals can be easily found even in places that are inaccessible to boats or carts. The value of top-grade gold is 16,000 times than that of black iron. However, if there are no iron boilers and iron axes used in daily lives even valuable gold will not be beneficial to human beings. The livelihood of all depends on the trading of what the people have for what they have not, as provided in the chapter on goods and exchange in

The Rites of Zhou. Then who was the first one to judge metals' qualities and their values which make them the necessities forever?

宋子曰,兵非圣人之得已也。虞舜在位五十载,而有苗犹弗率。明王圣帝,谁能去兵哉?"弧矢之利,以威天下",其来尚矣。为老氏者,有葛天之思焉,其词有曰:"佳兵者,不祥之器。"盖言慎也。火药机械之窍,其先凿自西番与南裔,而后乃及于中国,变幻百出,日盛月新。中国至今日,则即戎者以为第一义,岂其然哉!虽然,生人纵有巧思,乌能至此极也?

《佳兵第十五卷》

Songzi says that weapons are in use when wise men are at a corner. In ancient times Emperor Shun had been in power for 50 years and the Miao were not obedient to him. How can any of wise emperors ignore weapons? It has been passed on for a long time that weapons exists to terrorize the world. However Laozi was thought to agree with Getianshi's idea. In his book it was found that weapons were hoodoos, which indicates that weapons should be in use very cautiously. The technology to make gunpowder and firearms was derived from the Western world and South Asia. After it spread to China, it changed with each passing day. China ranks in the first place making powder and firearms. It might be true. Otherwise it cannot be so perfect if there are not a lot of efforts.

宋子曰,斯文千古之坠也,注玄尚白,其功孰与京哉?离火红而至黑孕其中,水银白而至红呈其变,造化炉锤,思议何所容也。五章遥降,朱临墨而大号彰。万卷横披,墨得朱而天章焕。文房异宝,珠玉何为?至画工肖像万物,或取本姿,或从配合,而色色咸备焉。夫亦依坎附离,而共呈五行变态,非至神孰能于斯哉?

《丹青第十六卷》

Songzi says that ancient cultural heritage will never perish from the earth due to the records by paper, brush pen and ink with an incomparable effect. The material for ink is in the black smoke from the burning of deal and tung oil. The material for writing and painting shall be obtained after the white mercury is burned and refined into red cinnabar. It is really amazing after the material is burned. The five color documents of state issued by the central government, make the significant orders delivered because the emperor signs the black characters of the order by using the brush pen with red ink. To take notes with red ink while reading books makes the excellent works more brilliant. These are the treasures that belong to a study where there is no room for pearls and gems. How can the bead or jade compare with them? The painters paint various paintings by using ink only or the mixture of cinnabar, ink and other dyes. The producing of cinnabar and other dyes must depend on the power of water and fire and reveal the changes of five elements. No one can fulfill this without the use of these natural resources in a flexible way.

宋子曰，狱讼日繁，酒流生祸，其源则何辜。祀天追远，沉吟《商颂》、《周雅》之间。若作酒醴之资曲蘖也，殆圣作而明述矣。唯是五谷菁华变幻，得水而凝，感风而化。供用岐黄者神其名，而坚固食馐者丹其色。君臣自古配合日新，眉寿介而宿痼怯，其功不可殚述。自非炎黄作祖，末流聪明，乌能竟其术哉！

《曲蘖第十七卷》

Songzi says that an increased number of crimes and litigations are caused by the bad effect of alcoholic drink on society. However, the curse does not lie in the brewing process itself. Good wine should be dedicated to God when a memorial ceremony is held. Wine is drunk to add to the fun when poems and music from

The Book of Poetry are enjoyed at ceremonies and banquets. As is clarified in the works of ancient oracles, yeast is indispensable to the brewing process. When making yeasts, the essence of the five grains is changed. After the grains are ground, they are mixed with water and kneaded into solid forms, then they become yeast by the breeze. The yeast used by physicians is called "medicinal yeast" and those red ones, which can make food delicious, are called "red yeast". Since ancient times, the blending methods of the principal materials and supplementary materials in making medicinal yeast have been improved. The functions of medicinal yeast in macrobiotics and in curing inveterate diseases are measureless. Without the great achievements accomplished by our ancestors Emperor Yan and Emperor Huang and the abilities and wisdom of the later generations, how else could this technique become so perfect?

宋子曰,玉蕴山辉,珠涵水媚,此理诚然乎哉,抑意逆之说也?大凡天地生物,光明者昏浊之反,滋润者枯涩之仇,贵在此则贱在彼矣。合浦、于阗行程相去二万里,珠雄于此,玉峙于彼,无胫而来,以宠爱人寰之中,而辉煌廊庙之上。使中华无端宝藏折节而推上坐焉。岂中国辉山媚水者萃在人身,而天地菁华只有此数哉?

《珠玉第十八卷》

Songzi says that the story goes that the mountains that produce jades are gleamy and the water that has pearls is radiant. Does this statement make sense? Or is it just a hypothesis? Everything in nature has its opposite, so that when something sparkles, there must be something dim, and when something is moist, there must be something dry. Though the distance between Hepu, the famous pearl-producing area, and Yutian, the area teaming with jades, is 20,000 li, these gems are purchased and cherished by people everywhere and even strived for their favors in the palace. Pearls and jades are considered as the primacy superior to other precious gems in China. However are these gems, adorning human bodies, the only treasures in China and thus the essence of excellence in the world?

拓展知识

下文是《齐民要术》的内容简介，由石声汉撰写。除文中的繁体字转换为简体字外，其余均为原文摘录。

The Book *Ch'i Min Yao Shu*

by Shih Shêng-Han

Among the now extant Chinese classics devoted solely to agriculture, 齐民要术 *Ch'i Min Yao Shu* (Essential Ways for Living of the Common People) is the best preserved and most comprehensive one. It was, as most investigators nowadays agree, written in the early 6th century A.D., most probably in the years 533 to 544. About its author 贾思勰 Chia Ssu-hsieh, nothing more is definitely known than that be was sometime a 太守 Tai Shou (governor, literally "Warden Major") of the District 高阳 Kaoyang, —this official title appears in the superscript on each of the ten fascicules of the various editions from the 宋 Sung period onwards.

Of this invaluable work, there are several editions available. Among them, the best and easiest to obtain is the 四部丛刊 Sze Pu Ts'ung Kan edition, a photocollographical impression based on a 明 Ming hand transcript which was reproduced from a 南宋 Later-Sung block-printing (an 1144 A.D. edition). Two earlier copies are kept in the libraries in Japan: one, a Sung impression from woodblocks made in 1023–1031, is a relic with only two complete fascicules and a few leaves of the first fascicule; the other, a transcript in scrolls (made by some Japanese buddhistic monks) of the fascicules 1, 2 and 4–10, probably based on a copy very similar to the former. None of these earlier editions is, however, free from misprints and faulty transcriptions, while later ones are definitely still worse. Now, however, a new revised edition with corrections (of misprints), annotation and transliteration into modern colloquial Chinese[①], is available.

Chia Ssu-hsieh himself summarized the contents and aim of his book in the Preface:

"今采捃经传,爰及歌谣,询之老成,验之行事。起自耕农,终于醯醢,资生之业,靡不毕书。号曰:《齐民要术》。凡九十二篇,分为十卷,……舍本逐末,贤哲所非;日富岁贫,饥寒之渐。故商贾之事,阙而不录。花草之流,可以悦目,徒有春花,而无秋实,匹诸浮伪,盖不足存。……"

"(So I) have made excerpts from classics, contemporary books, proverbs and folksongs, gathered informations from experts, and drawn from personal experience. Beginning with ploughing and cultivation, down to the making of vinegar and meat-pastes, any art useful in supporting the daily life (of the common folk) is jotted down. The book is thus named *Ch'i Min Yao Shu* (Essential Ways for Living of the Common People). The 92 Chapters are divided into 10 scrolls (volumes or fascicules). ...Sagacious man never will abandon the basic occupation and chase after the frivolous[②]; mending long straits with sudden profit usually brings in hunger and cold. Therefore no account is given for commercial dealings. Flowers may certainly be pleasant for the eye, but empty blooms in spring without

① Published by the Science Press, Peking, China, in 4 parts, 1957–1958.

② According to the then prevailing "Confucian" social teaching, farmers, being the second class, are the "basic" elements of the population, while merchants and pedlers being the fourth, i.e. the lowest class; chasing after mercenary gains ought to be despised as mean men.

substantial autumn fruits are vain and fraudulent things. So there is no need to record them. ..."

Hence, it is clear that the book is primarily intended as a practical guide for the improvement of rural life in general.

For a better apprehension of the scope of the *Ch'i Min Yao Shu*, it is necessary to list the main topics of the 92 chapters in the 10 fascicules:

Fasc. Ⅰ:

Chapter 1, reclamation of land; 2, seed-corn; 3, spiked millet.

Fasc. Ⅱ:

Chapter 4, panicled millets; 5, Setaria; 6, larger-grained beans, i.e. soya; 7, smaller-grained lesser beans; 8, hemp; 9, seed-hemp; 10, barley and wheat; 11, paddy rice; 12, land rice; 13, sesame; 14, melons; 15, gourds; 16, taro.

Fasc. Ⅲ:

Chapter 17, mallow; 18, turnips; 19, garlics; 20, scallions; 21, Allium fistulosum; 22, chives; 23, white mustard and other Brassica vegetables; 24, coriander; 25, basil; 26, Perilla and Polygonum; 27, ginger; 28, Zingiber mioga, celery, and Lactuca denticulata; 29, alfalfa; 30, "Journal" (a monthly calendar for farming house-economy).

Fasc. Ⅳ:

Chapter 31, living hedge; 32, transplantaton of trees; 33, jujube; 34, peaches; 35, gages; 36, plums and apricots; 37, grafting pear-trees; 38, chestnuts; 39, apples; 40, persimmons; 41, pomegranates; 42, Chaenomeles; 43, Zanthoxylon piperatum; 44, Zanthoxylon ailanthoides.

Fasc. Ⅴ:

Chapter 45, mulberry-trees and sericulture; 46, elm and poplar; 47, crab-apples; 48, Broussonetia; 49, care of lacquer-wares; 50, Sophora, Salix, Catalpa, Firmiana and Quercus; 51, bamboos; 52, bastard saffron; 53, indigo plant; 54, Lithospermum; 55, tree-felling.

Fasc. Ⅵ:

Chapter 56, care of cattle, horses, donkeys and mules; 57, sheep and goats; 58, swine; 59, poultry; 60, geese and ducks; 61, how to keep fish ponds.

Fasc. Ⅶ:

Chapter 62, produce and mercantile demand; 63, curing of earthen ware; 64, quick starters (for brewery) and wines; 65, white must starters; 66, heavy starters and wines; 67, formulary elixirs.

Fasc. Ⅷ:

Chapter 68, wheat-must and malt; 69, refining table salt; 70, chiang (protein-hydrolysates); 71, vinegars; 72, melanized soja; 73, a condiment (with 8 components); 74, cha (meat preserved in a medium of lactic fermentation); 75, bacon and jerked meat; 76, bouillons and stews; 77, steamed meat; 78, frying and boiling; 79, meat cooked with pickles and other sour preparations.

Fasc. Ⅸ:

Chapter 80, roasts; 81, meat-jellies; 82, wheat-flour meals; 83, dumplings; 84, porridges; 85, starch jellies; 86, rice-meals; 87, vegetarian dishes; 88, pickles and allied conserves; 89, candies; 90, glue-making; 91, manufacture of brush-pens and ink-sticks.

Fasc. Ⅹ:

Chapter 92, cereals, fruits, melons and other vegetable food not indigenous to Northern China—actually including those useful plants indigenous to but not cultivated in the Yellow River region, those useful but neither indigenous nor cultivated there, and those neither very useful nor indigenous nor cultivated.

Such an arrangement of subject matter in the first six fascicules strongly suggests an "artificial classification" of living beings. Thus, the objects of chapters 3–13 are what the ancient Chinese called "五谷" or "六谷" (five or six cereals); 14 and 15, pepos; 16, a plant very often used as staple food. Chapters 17–29 deal with various culinary vegetables; and these may be subdivided into 17, glossy, 18–23, "rank", 24–27, spices. Fruits are treated in fasc. Ⅳ in the order: 33–38, principal orchard fruits; 39–42, montanic fruits; 43 and 44, spicy fruits. Fascicule Ⅴ, deals with timber wood and tinctorial plants, and fascicule Ⅵ is alloted exclusively to domesticated animals in the order: mammals, birds and fishes. This system of classification is the basic form adopted in all "natural histories", herbalist writings and "materia medica" of China. Since, however, of the extant books on these subjects none is authentically earlier than the *Ch'i Min Yao Shu*, for the time being we have to credit Chia Ssu-hsieh with the priority of using such a system, anyhow in written form.

A glance at the table of contents of the *Ch'i Min Yao Shu* gives the impression that it actually covers all the general aspects of farming knowledge. Roughly speaking, the main body of informations in the first 9 fascicules falls into two categories: those on agrarian operations (namely Fasc. I to VI) and those on domestic occupations (Fasc. VII, VIII, IX and sundry sections in I to VI). It is the former 9 fascicules which interest us most, and now we shall proceed to see what can be made out of them.

练 习

1. Which field is *Ch'i Min Yao Shu* primarily dedicated to ?

   A. Philosophy.　　B. Agriculture.

   C. Literature.　　D. History.

2. Who is the author of *Ch'i Min Yao Shu*?

   A. Shih Shêng-Han.　　B. Chia Ssu-hsieh.

   C. Tai Shou.　　D. Kaoyang.

3. When was *Ch'i Min Yao Shu* most likely written?

   A. 5th century A.D.

   B. 6th century A.D.

   C. 7th century A.D.

   D. 8th century A.D.

4. Which edition of *Ch'i Min Yao Shu* is considered the best and easiest to obtain?

   A. Sung edition.

   B. Ming edition.

   C. Sze Pu Ts'ung Kan edition.

   D. Science Press edition.

5. What is the main purpose of *Ch'i Min Yao Shu*?

   A. Record commercial dealings.

   B. Provide information on flowers and plants.

   C. Serve as a practical guide for rural life improvement.

   D. Analyze social classes in ancient China.

6. In which district was Chia Ssu-hsieh a former governor?

A. Kaoyang.　　B. Sung.

C. Ming.　　D. Peking.

7. Which country's libraries hold earlier copies of *Ch'i Min Yao Shu*?

A. India.　　B. Japan.

C. Korea.　　D. Vietnam.

8. When was the Science Press edition of *Ch'i Min Yao Shu* published?

A. 1930—1931.　　B. 1957—1958.

C. 1980—1981.　　D. 2000—2001.

9. What topics are covered in the first six fascicules of *Ch'i Min Yao Shu*?

A. Culinary arts.　　B. Domestic animals.

C. Timber wood.　　D. Agrarian operations.

第八章　农业科技典籍语篇的翻译

◎本章学习目标◎

1. 了解农业科技典籍语篇的特点
2. 理解衔接和连贯的含义
3. 理解照应、重复、连接的含义和原则
4. 掌握农业科技典籍语篇的翻译方法

衔接(cohesion)与连贯(coherence)是构建语篇的重要因素。衔接是连贯的外在形式,连贯是衔接的内在意义。[①]英语是形合语言,语义上的连贯往往通过显性衔接达成,而汉语是意合语言,多用隐性衔接实现语义连贯。在农业科技典籍的语篇翻译中,需要利用多种衔接手段,显化隐含的衔接关系,使译文更符合英语语篇的组织架构,在语义上是一个能被读者理解的连贯整体。

第一节　照　应

【例1】

稻人掌稼下地。以“猪”蓄水,以“防”止水,以“沟”荡水,以“遂”均水,以

① 何善芬.英汉语言对比研究[M].上海:上海外语教育出版社,2002:466-477.

"列"舍水,以"浍"写水,以涉扬其芟,作田。

The Tao Jen or Rice beadle: His duty is to cultivate low lands. He deposits water in reservoirs, stops it with dykes, passes it on with channels, distributes with branch channels, feeds with furrows, and leads it away with ditches. Then, men can wade through the standing water and clear away weeds formerly hewn and cultivate.

(《周礼》,石声汉 译)

原文只出现了一个主语"稻人",第一句的意思是:稻人掌管在低地种庄稼。第二句的译文增加了人称代词He,表明这里列举的是稻人的六项具体工作,与第一句衔接更紧密。最后的"以涉扬其芟,作田"不再是稻人专属职责,众人皆可参与,译文增加了主语men。补充的两个主语使得整篇的语义更明确。类似的例子还有:

【例2】

荞麦之茎弱而翘然,易长易收,磨面如麦,故曰荞曰荍,而与麦同名也。俗亦呼为甜荞,以别苦荞。杨慎丹铅录,指乌麦为燕麦,盖未读日用本草也。

Qiaomai has weak, but upright stem. It is a herb that grows easily and can be harvested conveniently. The seed can be ground into flour, which is similar to Xiaomai / semen tritici / wheat. It is colloquially called Tianqiao (meaning "sweet Qiao") to differentiate from Kuqiao-mai (meaning "bitter Qiao"). Yang Shen in his work *Danqian Lu* said that Wumai was just Yanmai (Quemai/herba bromi japonica / herb of Japanese bromegrass), but he was wrong. He did not read of the book *Riyong Bencao* (*Materia Medica for Daily Use*).

(《本草纲目》,罗希文 译)

【例3】

人之情不能无衣食,衣食之道,必始于耕织,万民之所公见也。物之若耕织者,始初甚劳,终必利也。

It is an essential quality of human beings that they cannot do without clothing and food. The Way of clothing and feeding oneself must begin with farming and weaving. This is something that the people in their tens of thousands all recognize. Things like farming and weaving begin with hard work, but in the end they are inevitably beneficial.

(《淮南子》,John S. Major等 译)

【例4】

陕西因洪水下大石,塞山涧中,水遂横流为害。石之大有如屋者,人力不能去,州县患之。雷简夫为县令,乃使人各于石下穿一穴,度如石大,挽石入穴窖之,水患遂息也。

Mountain torrents rushed down in Shaanxi and huge rocks blocked the river course in the mountain. As a result, the river was blocked and started to inundate. As some rocks were as huge as the size of a house, people were unable to move them. At that time Lei Jianfu was the county magistrate. He sent people to estimate the size of the rocks before digging a pit beneath each of them. This being done, the rocks were pushed into the pits one by one. In such a way the problem caused by mountain torrents was solved.

(《梦溪笔谈·权智卷十三》,王宏、赵峥 译)

原文记述了陕西因爆发山洪落下大石块引发水患的事件。译文中增补了三个状语,at that time照应前文事件时间,this being done照应前文雷简夫的做法,in such a way照应前文推石入坑止水患的办法。类似的例子还有:

【例5】

种棉有漫种者,易种难锄,穴种者反之。漫种者,下种宜密;锄时,简别而痛芟之,令绝疏。

To sow cotton broadcast is easy, but it is troublesome to weed it, while the contrary is the case in dibbling it; in the former way, too, the plants easily grow too thick, and weeding and thinning them is much more troublesome.

（《农政全书》卷三十五《蚕桑广类·木棉》，C. Shaw 译）

练　习

1. 翻译以下典籍原文，特别注意上下文照应。

1）草人，掌土化之法。以物、地，相其宜而为之种。（《周礼》）

2）贤者与民并耕而食，饔飧而治。（《孟子》）

3）养羊法，当以瓦器盛一升盐，悬羊栏中。羊喜盐，自数还啖之，不劳人牧。羊有病，辄相污。欲令别病法：当栏前作渎，深二尺，广四尺。往还皆跳过者，无病；不能过者，入渎中行；过，便别之。（《家政法》）

4）宋明帝好食蜜渍鱁鮧，一食数升。鱁鮧乃今之乌贼肠也，如何以蜜渍食之？大业中，吴郡贡蜜蟹二千头，蜜拥剑四瓮。又何胤嗜糖蟹。大抵南人嗜咸，北人嗜甘。鱼蟹加糖蜜，盖便于北俗也。如今之北方人喜用麻油煎物，不问何物皆用油煎。庆历中群学士会于玉堂，使人置得生蛤蜊一篑，令饔人烹之。久且不至，客讶之，使人检视，则曰："煎之已焦黑而尚未烂。"坐客莫不大笑。予尝过亲家设馔，有油煎法鱼，鳞鬣虬然，无下箸处，主人则捧而横啮，终不能咀嚼而罢。（《梦溪笔谈》）

5）草壅甚热，过于粪饼。粪因水解，饼亦匀细。草壅难匀，当其多处，峻热伤苗，故有时倍收，有时耗损。用此一物，特宜详慎。（《农政全书》）

6）（荞麦）酸，微寒，食之难消。久食动风，令人头眩。作面和猪、羊肉热食，不过八九顿，即患热风，须眉脱落，还生亦希。泾、邠以北，多此疾。又不可合黄鱼食。（《本草纲目》）

7）（荞麦）北方多种。磨而为面，作煎饼，配蒜食。或作汤饼，谓之河漏，以供

常食,滑细如粉,亚于麦面。南方亦种,但作粉饵食,乃农家居冬谷也。(《本草纲目》)

8)得之必生,失之必死者,何也?唯粟。得之,尧舜禹汤文武孝已,斯待以成,天下必待以生。故先王重之。一日不食,比岁歉;三日不食,比岁饥;五日不食,比岁荒;七日不食,无国土;十日不食,无畴类,尽死矣。(《管子》)

2.将以下两段话译成英文。

1)《周易》被誉为诸经之首,是阐述天地世间万象变化的古老经典,是博大精深的辩证法哲学书。其内容包括《经》和《传》两个部分,《经》主要是六十四卦和三百八十四爻,卦和爻各有说明,作为占卜之用。《传》包含解释卦辞和爻辞的七种文辞共十篇,统称《十翼》,相传为孔子所撰。它的内容极其丰富,包罗万象,对中国几千年来的政治、经济、文化等各个领域都产生了极其深刻的影响。

2)在文字学意义上,"年"的本义指庄稼成熟,即年成。因庄稼大都一岁一熟,"年"渐等同于"岁",成为历法上的时间单位(一年),后又引申指年节(春节)。在历法意义上,它是指中国传统农历(阴阳合历)的一个时间周期,平年12个月,大月30天,小月29天,全年354或355天;闰年13个月,全年383、384或385天。作为一个时间周期,它与中国古代的农业生产密切相关,反映农耕社会的时间意识和思想观念。近代以来,西方的历法(公历)传入中国,1912年为中华民国正式采用,形成了公历与农历并行的双历法系统,所以"年"现在既指农历的时间周期,也指公历的时间周期,视具体的语境而定。

3.下文记述了在播种前处理种子的方法,根据释文进行翻译。

取马骨,剉,一石以水三石煮之。三沸,漉去滓,以汁渍附子五枚。三四日,去附子,以汁和蚕矢羊矢各等分,挠,令洞洞如稠粥。先种二十日时,以溲种，如麦饭状。当天旱燥时，溲之，立干；薄布，数挠，令易干；明日复溲。天阴雨则勿溲！六七溲而止。辄曝，谨藏；勿令复湿。至可种时，以余汁溲而种之，则禾稼不蝗、虫。

无马骨，亦可用雪汁。雪汁者，五谷之精也，使稼耐旱。常以冬藏雪汁，器盛，埋于地中。治种如此，则收常倍。

（《氾胜之书》）

【释文】

将马骨，斫碎，一石碎骨，用三石水来煮。煮沸三次后，过滤掉骨渣，把五个附子泡在清汁里。三四天以后，又漉掉附子，把分量彼此相等的蚕粪和羊粪加下去，搅匀，让混合物像稠粥一样地稠。下种前二十天，把种子在这糊糊里拌和，让每颗种子都黏上一层糊糊，结果变成和麦饭一样。一般只在天旱、空气干燥时拌和，所以干得很快；再薄薄地铺开，再三拌动，叫它更容易干；第二天，再拌再晾。阴天下雨就不要拌。拌过六七遍，就停止。立刻晒干，好好地保藏；不要让它潮湿。到要下种时，将剩下的糊糊再拌一遍后再播种，这样的庄稼，不惹蝗虫和其他虫害。

没有马骨，也可用雪水代替。雪水是五谷之精，可以使庄稼耐旱。记住经常在冬天收存雪水，用容器保存，埋在地里准备着。这样处理种子，常常可以得到加一倍的收成。

延伸阅读

《吕氏春秋》译文赏析

《吕氏春秋》共有三个英译本，一个是诺布诺克和里格尔的合译本（以下简称“合译”），一个是翟江月译本（以下简称“翟译”），一个是汤博文译本（以下简称“汤译”）。

下文选自《吕氏春秋·孟春纪第一·孟春》，记述了早春时节草木萌动，农事初始的情景。

是月也,天气下降,地气上腾,天地和同,草木繁动。王布农事,命田舍东郊,皆修封疆,审端径术。善相丘陵阪险原隰,土地所宜,五谷所殖,以教道民,以躬亲之。田事既饬,先定准直,农乃不惑。

合译:In this month, Celestial ethers descend, Terrestrial ethers ascend, Heaven and Earth harmoniously unite. And grasses and trees begin to sprout and grow. The king distributes the tasks of agriculture and commands that field inspectors lodge at the eastern suburban altar. They are to insure that everyone keeps boundaries and borders in good repair and that care is taken as to the straightness of the small pathways between fields. They are skillfully to survey the mounds, slopes, ravines, plains, and marshes to determine which have soil and landforms suitable to grow each of the five grains. In all this they must instruct the people and personally participate in the work. When before tasks in the fields are announced, the boundaries have all been fixed, the farmers will harbor no suspicions.

翟译:During this month, the vitality of Heaven descends, and on the other hand, the vitality of Earth ascends. These two kinds of vitalities combine together. As a result, plants start to germinate. The Son of Heaven issues edicts on farm work. Sub-officials in charge of agricultural affairs are sent to the eastern suburb to supervise people in cultivating fields and maintaining paths. They are to investigate the conditions of the land, such as hills, mountainous areas, plains and the low-lying areas, and then tell farmers which kind of crops should be grown on which fields. They should also perform the farm work in person to show farmers the correct way to do it. After that, a set of standards are set up, so that farmer will not be confused.

汤译：In this month, the vapor of heaven moves downward and that of the earth moves upward. The merging of heaven and earth brings the grass and trees to luxuriant growth. The king is to make arrangements for farming. He is to order the official in charge of farming to stay in the eastern outskirts to supervise the repair of boundaries, examine and straighten the paths in the fields, study the hills, highlands, plains and marshes and see what crops are suitable to the land and what land is suitable to the crops and then teach the people accordingly. He must see to everything in person. When arrangements for farming are made and standards set, the farmers will know exactly what to do.

第二节　重　复

【例1】

水之性，行至曲，必留退，满则后推前，地下则平行，地高即控，杜曲则捣毁，杜曲激则跃，跃则倚，倚则环，环则中，中则涵，涵则塞，塞则移，移则控，控则水妄行，水妄行则伤人，伤人则困，困则轻法，轻法则难治，难治则不孝，不孝则不臣矣。

According to the nature of water, it will stop flowing and recede when it reaches extremely steep places. If a place is full of water, the water behind will drive the water in front of it and advance meanderingly. It will progress smoothly when it encounters lower-lying places. And it will splash about when a higher place is encountered. If the surface of the ground is winding, it will cause a landslide. And if the surface is too rough, it will skip over some places. When it skips over some places, it will swirl. When it swirls, there will be whirlpools. Where there are whirlpools, it will accumulate at some places. Where it accumulates, mud and sand it carried will deposit. When mud and sand deposits, the water course will be silted. Where it is silted, it will change its original course. When it changes course, it will surge. When it surges, it will flow unrestrained. When it is unrestrained, it will harm people. When the people are harmed by flood, they will be impoverished. When they are impoverished, they will not attach importance to the law. If they do not attach importance to the law, it will be difficult to govern them. If they cannot be governed, there will be no filial piety. If they have no filial piety, they will not obey the sovereign.

(《管子·度地第五十七》,翟江月 译)

前后文首尾重复是文言文中的常见句型,也是一种修辞手法"顶真",翻译时可保留重复。原文记述了水的特点以及水害的危险,文中有大量的首尾重复,比如如果地势过于曲折,水流就会跳跃;跳跃就会溢出;溢出就形成漩涡……。译文保留了这些上下文的因果逻辑衔接,语义流畅连贯。类似的例子还有:

【例2】

帝俊生三身,三身生义均,义均是始为巧倕,是始作下民百巧。后稷是播百谷。稷之孙曰叔均,是始作牛耕。

King Jun gave birth to Sanshen. Sanshen gave birth to Yijun. Yijun was also called Qiaochui. Qiaochui taught people a hundred crafts. Houji sowed a hundred grains. The grandsons of Houji was Shujun, who was the first one to plough farm fields with oxen.

(《山海经·海内经》,王宏、赵峥 译)

【例3】

东方生风,风生木,木生酸,酸生肝,肝生筋,筋生心,肝主目。

The east produces wind, the wind promotes the growth of trees, the trees produces sour taste, the sour taste nourishes the liver, the blood stored in the liver nourishes the sinews, the sinews nourishes the heart and the liver controls the eyes.

(《黄帝内经·素问》,李照国 译)

除了前后文首尾重复,原文中重复的关键词或句子结构也可以在译文中保留。如下面两例中,"守"在译文统一处理为safeguard;"而能""而"句子结构也在译文中用but able to和and进行重复。

【例4】

地之守在城,城之守在兵,兵之守在人,人之守在粟。

Circumvallation is the safeguard to provide safety of the territory of the state; troops are the safeguard to provide safety of the protective walls; the safeguard for the troops is the common people; and the safeguard for the common people is grain.

(《管子·权修第三》,翟江月 译)

【例5】

约而能张,幽而能明,弱而能强,柔而能刚,横四维而含阴阳,纮宇宙而章三光。

It is constrained but able to extend. It is dark but able to brighten. It is supple but able to strengthen. It is pliant but able to become firm. It stretches out the four binding cords and restrains yin and yang. It suspends the cosmic rafters and displays the Three Luminaries.

(《淮南子》,John S. Major等 译)

此外,为了明确或强调,即使原文没有重复,翻译时有必要对一些内容进行重复。比如:

【例6】

《庄子》曰:“畜虎者不与全物、生物。”此为诚言。尝有人善调山鹧使之斗,莫可与敌。人有得其术者,每食则以山鹧皮裹肉哺之,久之,望见真鹧,则欲搏而食之。此以所养移其性也。

Zhuangzi says, “The man who raises a tiger does not feed it with a whole animal or a living animal.” This is quite right. There was a man who was good at training partridges. When his partridge was put into a fight, no other partridges could defeat it. Later his training method was known to others. The man always fed his partridge with meat covered with the skin of another partridge. Being trained in such a way for a long time, his partridge would attack and devour areal one at the sight of it. This shows how the way of feeding can change the habits and characteristics of the partridge.

(《梦溪笔谈·权智卷十三》,王宏、赵峥 译)

原文讲述了喂养条件改变动物生活习性的道理,文中“山鹧”出现了三次,而译文中partridge一词出现了七次,用同词重复加强衔接和连贯。

练 习

1.翻译以下典籍原文,特别注意重复的使用。

1)区中草生,茇之。区间草,以利剗剗之,若以锄锄。(《氾胜之书》)

2)一年之计,莫如树谷;十年之计,莫如树木;终身之计,莫如树人。一树一获者,谷也,一树十获者,木也,一树百获者,人也。(《管子》)

3)民事农则田垦,田垦则粟多,粟多则国富,国富者兵强,兵强者战胜,战胜者地广。是以先王知众民、强兵、广地、富国之必生于粟也。(《管子》)

4)凡治国之道,必先富民,民富则易治也,民贫则难治也。奚以知其然也?民富则安乡重家;安乡重家,则敬上畏罪;敬上畏罪,则易治也。民贫则危乡轻家；危乡轻家，则敢陵上犯禁；陵上犯禁，则难治也。故治国常富而乱国常贫。是以善为国者，必先富民，然后治之。(《管子》)

5)南方生热，热生火，火生苦，苦生心，心生血，血生脾，心主舌。(《黄帝内经》)

6)中央生湿，湿生土，土生甘，甘生脾，脾生肉，肉生肺，脾主口。(《黄帝内经》)

7)西方生燥，燥生金，金生辛，辛生肺，肺生皮毛，皮毛生肾，肺主鼻。(《黄帝内经》)

8)北方生寒，寒生水，水生咸，咸生肾，肾生骨髓，髓生肝，肾主耳。(《黄帝内经》)

9)食者，民之本也；民者，国之本也；国者，君之本也。(《淮南子》)

10)得以利者不能誉，用而败者不能非，收聚畜积而不加富，布施禀授而不益贫，旋县而不可究，纤微而不可勤，累之而不高，堕之而不下，益之而不众，损之而不寡，斲之而不薄，杀之而不残，凿之而不深，填之而不浅。(《淮南子》)

2.将以下两段话译成英文。

1)17世纪，欧洲进入历史大变革时代，资本主义正在兴起；在中国，资本主义也开始萌芽，农业和手工业等生产技术全面提高。在这个集传统科技之大成而又充满新气息的时代，明朝中后期出现了一批反映中国古代科技成就的巨著，宋应星著的《天工开物》便是其中的杰出代表。

2)《韩非子》由五十五篇独立成篇的文章结集而成，除个别篇外，篇名皆表明该篇主旨。全书主要阐述了韩非以君主专制主义为基础的法、术、势相结合的法治思想和主张，以及他的进化历史观和讲求实际的哲学思想。韩非在论述其学术思想时大量征引春秋战国时期的故事、传说、历史事件、人物言论等，这些材料也成为研究这一时期历史的宝贵史料。

3.下文记述了粮食生产和农业发展对于治国的重要性,根据释文进行翻译。

不生粟之国亡,粟生而死者霸,粟生而不死者王。粟也者,民之所归也;粟也者,财之所归也;粟也者,地之所归也。粟多则天下之物尽至矣。故舜一徙成邑,二徙成都,参徙成国。舜非严刑罚、重禁令而民归之矣,去者必害,从者必利也。先王者,善为民除害兴利,故天下之民归之。所谓兴利者,利农事也。所谓除害者,禁害农事也。农事胜则入粟多,入粟多则国富,国富则安乡重家,安乡重家则虽变俗易习,敺众移民至于杀之,而民不恶也,此务粟之功也。上不利农则粟少,粟少则人贫,人贫则轻家,轻家则易去,易去则上令不能必行,上令不能必行则禁不能必止,禁不能必止则战不必胜,守不必固矣。夫令不必行,禁不必止,战不必胜,守不必固,命之曰寄生之君。此由不利农少粟之害也。

粟者,王之本事也,人主之大务。有人之涂,治国之道也。

(《管子·治国第四十八卷》)

【释文】

一个国家不生产粮食,就会灭亡;能生产粮食但当年就吃光用尽的国家仅能称霸;生产粮食但却食用不尽的国家能成就王业。粮食能吸引人民,粮食能招引财富,粮食还能拓展疆域。一个国家的粮食多了,天下所有的物产都会被吸引过来。所以,舜第一次迁徙建成一个“邑”,第二次迁徙建成一个“都”,第三次迁徙建成一个“国”。舜并没有采用严苛的刑罚和禁令,但人民都跟定了他,因为离开他,人们就会受害;而跟随他,人们就会受益。古代圣明的君主善于为人民兴利除害,所以天下人都归附他们。所谓兴利,就是采取行动以促进农业生产。所谓除害,就是禁止妨害农事。农业发展了,粮食就会丰足;粮食丰足,国家就会富庶;国家富庶,人民就会安居乐业并且珍惜自己的家园;人民安居乐业、珍惜自己的家园,即使改变他们的风俗习惯,对他们进行役使,甚至有所杀戮,他们都不会对君主心存怨恨。这都是致力于粮食生产的功效。君主不采取措施促进农业生产的发展，粮食必然会少；粮食少了，人民就会贫困；人民贫困，他们就会轻视家园；轻视家园，就容易外逃；人民外逃，所有的命令就不能完全被执行；命令不能完全执行，禁律也不能完全制止；禁律不能完全制止，战争就不一定必胜，防守也不一定必固。命令不能必行，禁律不能必止，出战不能必胜，防守不能必固，这种情况下，国君就叫“寄生的君主”。这都是不促进农业生产导致缺少粮食而造成的危害。

所以粮食是成就王业的根本，粮食生产是君主的重大任务，还是吸引民众的途径和治理国家的道路。

延伸阅读

《吕氏春秋》译文赏析

《吕氏春秋》共有三个英译本，一个是诺布诺克和里格尔的合译本（以下简称“合译”），一个是翟江月译本（以下简称“翟译”），一个是汤博文译本（以下简称“汤译”）。

下文选自《吕氏春秋·士容论第六·上农》，记述了农业在陶冶心志、影响民风方面所发挥的作用。

古先圣王之所以导其民者，先务于农。民农非徒为地利也，贵其志也。民农则朴，朴则易用，易用则边境安，主位尊。民农则重，重则少私义，少私义则公法立，力专一。民农则其产复，其产复则重徙，重徙则死处而无二虑。舍本而事末则不令，不令则不可以守，不可以战。民舍本而事末则其产约，其产约则轻迁徙，轻迁徙，则国家有患，皆有远志，无有居心。民舍本而事末则好智，好智则多诈，多诈则巧法令，以是为非，以非为是。

合译：Of the methods used by the sage-kings of antiquity to guide their people, the first in importance was devotion to farming. The people were made to farm not only so that the earth would yield benefits, but also to ennoble their goals. When the people farm, they remain simple, and being simple are easy to use. Being easy to use, the borders are secure, and the position of ruler is honored. When the people farm, they are serious and hence seldom hold personal moral beliefs. When they seldom hold personal moral beliefs, then the law common to everyone is firmly established and all efforts are united. When the people farm, their household income increases, and when income increases, they are reluctant to move away. When they are reluctant to move away, they will spend their whole lives in their home villages and will not consider any other occupation. When the people abandon the fundamental occupation to pursue a secondary task, they will not obey orders. If the people do not obey orders, it will be impossible to use them, either to defend the state or wage war. When the people abandon the fundamental and pursue the secondary, their household income is meager, and when income is meager they will think nothing of moving away. When the people think nothing of moving, then should the nation face some difficulty, everyone will be thinking of how to flee to some distant place and no one will be of a mind to remain. When the people abandon the fundamental to pursue the secondary, they become fond of using their wits, and because they are fond of using their wits, their scheming increases. When scheming increases, the people try to outsmart laws and orders by making right into wrong and wrong into right.

翟译: The first important thing for the sage sovereigns of ancient times was to lead the common people to focus on farming. Leading the common people to focus on farming, they not only aimed at seeking the profits provided by the products of the fields, but also wanted to edify the will of their people this way, and the latter was much more important. When the common people are engaged in farming, they will become simple. When they are simple, it will be easy for the sovereign to use them. When it is easy to use them, the safety of the border areas will be secured, and the sovereign will become powerful and honourable. When the common people are engaged in farming, they will become sober. When they are sober, they will seldom discuss government affairs in private. When they seldom discuss government affairs in private, the legal system of the state will become well established, and the financial resources of the people will be easily operated. When the common people are engaged in farming, they will accumulate more possessions. When their possessions are numerous, they will not want to move to other areas. When they do not want to move to other areas, they will stay till the end of their lives at their hometowns and will not think of betraying the regime. If the common people give up farming to take up industry and commerce, they will not obey the sovereign's orders. When they do not obey the orders, they cannot be used to defend the state or attack other states. When the common people give up farming to take up industry and commerce, they will have less possessions. When they have less possessions, they will move to other areas at random. When they don't regard moving to other areas as a serious thing, they will want to leave their state when it is in danger and won't feel settled down in any case. When the common people give up farming to take up industry and commerce, they will use petty tricks. When they use petty tricks, they will become deceitful. When they are deceitful, they will misinterpret the law and edicts on purpose. Thus they will take wrong for right and take right for wrong.

汤译：The way in which the sage kings of the past gave guidance to the people was laying emphasis, first of all, on farming. Engaging the people in farming is not only for the purpose of obtaining yields from the land, but also for the purpose of improving their character. When the people are engaged in farming, they will become simple and honest. When they are simple and honest, they can be easily recruited for service. When they can be easily recruited for service, the frontiers will be secure and the ruler's position elevated. When the people are engaged in farming, they will become prudent. When they are prudent, they will talk less critically in private. When they talk less critically in private, law and order can be maintained and their strength can be concentrated. When the people are engaged in farming, their property will multiply. When their property has multiplied, they will be afraid to move. When they are afraid to move, they will stay at their home villages until they die without entertaining other thoughts. If the people abandon farming, which is the root, and are engaged in crafts and trade, which are the branches, they will become disobedient to orders. When they are disobedient to orders, they cannot be relied upon to defend and attack. If the people abandon agriculture and are engaged in crafts and trade, their property will become movable. When their property is movable, they can conveniently move to other places. When they can conveniently move to other places, they will move away and will not stay in the same place when the country is invaded by an enemy. If the people abandon farming and are engaged in crafts and trade, they will become more calculating. When they become more calculating, they will become crafty in many ways. When they become crafty in many ways, they will try to evade the law, turning right into wrong and wrong into right.

第三节　连　接

衔接的方法之一是增加连接(conjunction),使译文形成一个有机整体,在语义上能被读者理解。如《本草纲目》中记述的将荞麦作为食物的方法:荞麦作饭,须蒸使气馏,烈日暴令开口,舂取米仁作之。罗希文译本是这样处理的:First steam buckwheat and then dry it in the sun until it cracks. Husk the thing to get the grain, and then it can be used to make food. 译文增补了表示先后顺序的连接词,使得荞麦加工的步骤和过程更加清晰明了。

【例 1】

凡种谷,雨后为佳:遇小雨,宜接湿种;遇大雨,待萝生。小雨,不接湿,无以生禾苗;大雨,不待白背,湿辗,则令苗瘦。薉若盛者,先锄一遍,然后纳种,乃佳也。

In sowing spiked millets, it is better to follow a rain. If it is a light drizzle, catch up the damp; if a heavy shower, wait till weeds sprout. For, in drizzles, the seed corn would not sprout unless sown with enough wetness; after a heavy rain, when the surface was not yet tolerably dry to appear pale, treading will make the ground stickly and the seedlings will be weak. If weeds should grow to unwonted height, hoe them down before sowing.

(《齐民要术》卷一《种谷第三》,石声汉 译)

"薉"(huì),杂草,亦作"秽"。原文的意思是:种谷子,都以雨后下种为好:下过小雨,趁湿时种;下大雨,等杂草发芽后再种。雨下得小,不趁湿时下种,禾苗不容易生出;雨大了,如不等地面发白,湿着就去辊压,禾苗会瘦弱。杂草如果很多,先锄一遍,然后下种,才合适。译文增加了 if, for, unless, after, when 等连词,将语义中暗含的条件、原因、时间等关系显性化,

符合英语语篇的组织规范。类似的例子还有:

【例2】

麦生根成,锄区间秋草。缘以棘柴律土,壅麦根。秋旱,则以桑落时浇之。秋,雨泽适,勿浇之。春冻解,棘柴律之,突绝去其枯叶。区间草生,锄之。

After germination and well rooting, hoe to rid autumn-sprouting weeds. Drag thorn bunches through the rows to bank up wheat root. If the autumn is particularly dry, water the pits by the time of leaf-fall of mulberry trees. With timely autumn rainfall and proper soil-moisture, better not to water. With thawing in the spring, drag thorn bunches to detach dry dead leaves. Hoe to deweed amongst the pits.

(《汜胜之书》,石声汉 译)

【例3】

故圣人之处国者,必于不倾之地,而择地形之肥饶者,乡山左右,经水若泽,内为落渠之写,因大川而注焉。乃以其天材、地之所生,利养其人,以育六畜。天下之人皆归其德而惠其义。

And sages always choose to set up their capitals at secure places. Moreover, the fields there must be fertile; there must be mountains on both the right side and the left, and there should also be waters flowing across these places as well. They build canals inside the capital and lead into them the waters running in big rivers nearby. And then they can use the natural resources and grain produced on the land to support and benefit their people, and to help increase their livestock. Thus, people all over the world will admire their virtue and righteousness.

(《管子·度地第五十七》,翟江月 译)

原文的意思是:圣人建设都城,一定选在稳固的地方,而且土地必须肥沃，左右两面环山，中间还有河流或者湖泽经过，城内要修砌完备的沟渠排水，让水汇入大河。然后就可以利用自然资源和农业产品来供养国内人民，并促进六畜的繁育。这样，普天之下的人们都可以受到他们的德惠。

原文没有出现连接词，第一句话中间的递进、并列、条件等关系分别用moreover、and、then等词体现出来。两句话之间用thus进行连接，表明因果关系。另外，译文还增加了主语they指代圣人，形成上下文的照应。

【例4】

庆历中，议弛茶盐之禁及减商税。范文正以为不可：茶盐商税之入，但分减商贾之利耳，行于商贾未甚有害也；今国用未减，岁入不可阙，既不取之于山泽及商贾，须取之于农。与其害农，孰若取之于商贾？今为计莫若先省国用；国用有余，当先宽赋役，然后及商贾。弛禁非所当先也。其议遂寝。

During Qingli period of the reign of Emperor Renzong, there were talks of lifting the ban on the private trade of tea and salt and cutting down the taxes on commerce. Fan Zhongyan did not agree. He believed that the taxes collected from them may decrease the merchants' incomes, but could not bring too much damage to them. As the government budget did not decrease, its yearly income could not be cut down. If the money could not be collected from the trade in tea and salt and from commerce, the government will have to levy it on farmers. It was better to collect taxes from merchants than to harm the interests of farmers. Currently the best way is to cut down the government expenditure. Once the government budget had a favorable surplus, the top priority should be laid on cutting down the taxation on farmers and alleviating the heavy labour. As Fan did not agree to lifting the ban on the private trade of tea and salt and cutting down the taxes on commerce, talks of this kind died away.

（《梦溪笔谈·官政卷十二》，王宏、赵峥 译）

原文记述了范仲淹所采取的抑商重农的措施。译文中增补了表示转折(but)、原因(as)和条件(if, once)的连词,还增补了主语(he, Fan),加强上下文之间的照应。类似的例子还有:

【例5】

花性忌燥,燥则湿烝而桃易脱落。花忌苗并,并则直起而无旁枝,中下少桃。种不宜晚,晚则秋寒。早则桃多不成实,即成亦不甚大,而花软无绒。去心不宜于雨暗日,雨暗去心,则灌聋而多空干。此北方种花法也。北方地高寒,尚宜若此;况此中地湿燥,何可不以北法行之?

The cotton flower dreads hot weather, when the moist exhalations steam it, and cause the stamens to fall off easily. It also dislikes to have plants touch each other, for then the branches can not spread out well, and the flowers in the middle and bottom will be few. Cotton should not be planted too late, lest in the autumn the air be too cold for it; while if put in the ground too early, its blossoms will not set; or if they do, they will be small, and the flowers weakly, without cotton. Better not nip off the tops in dull and rainy weather, for then they will mostly get wet, and many of the branches be empty of flowers. These rules of growing cotton are adopted in the north; and if they are adopted in cold high regions, how much more would we here, where the land is damp and warm, find it advantageous to practice them?

(《农政全书》卷三十五《蚕桑广类·木棉》, C. Shaw 译)

练习

1. 翻译以下典籍原文,特别注意上下文的连接。

1)区田,以粪气为美,非必须良田也。诸山、陵,近邑高危,倾坂,及丘,城上,皆可为区田。区田,不耕旁地,庶尽地力。(《氾胜之书》)

2)今并州无大蒜,朝歌取种,一岁之后,还成百子蒜矣;其瓣粗细,正与条中子同。芜菁根其大如碗口,虽种他州子,一年亦变。大蒜瓣变小,芜菁根变大,二事相反,其理难推。又八月中方得熟,九月中始刈得花子。至于五谷蔬果,与余州早晚不殊,亦一异也。并州豌豆,度井陉已东,山东谷子入壶关、上党,苗而无实,皆余目所亲见,非信传疑。盖土地之异者也。(《齐民要术》)

3)凡栽一切树木,欲记其阴阳,不令转易。阴阳易位则难生。小小栽者,不烦记也。大树髡之;不髡风摇则死。小则不髡。先为深坑。内树讫,以水沃之,著土令如薄泥;东西南北,摇之良久,摇,则泥入根间,无不活者;不摇,根虚多死。其小树,则不烦耳。然后下土坚筑。近上三寸不筑,取其柔润也。时时溉灌,常令润泽。每浇,水尽即以燥土覆之。覆则保泽,不然则干涸。埋之欲深,勿令挠动。凡栽树讫,皆不用手捉及六畜抵突。(《齐民要术》)

4)柰、林檎,不种,但栽之。种之虽生,而味不佳。取栽,如压桑法。此果根不浮薉,栽故难求,是以须压也。又法:于树旁数尺许,掘坑,泄其根头,则生栽矣。凡树栽者皆然矣。(《齐民要术》)

5)凡伐木,四月、七月,则不虫而坚肕。榆荚下,桑椹落,亦其時也。然则凡木有子实者,候其子实将熟,皆其时也。非时者,虫而且脆也。凡非时之木,水沤一月,或火煏取干,虫皆不生。水浸之木,更益柔肕。(《齐民要术》)

6)忠定张尚书曾令鄂州崇阳县,崇阳多旷土,民不务耕织,唯以植茶为业。忠定令民伐去茶园,诱之使种桑麻,自此茶园渐少,而桑、麻特盛于鄂、岳之间。至嘉祐中,改茶法。湖、湘之民,苦于茶租,独崇阳茶租最少,民监他邑,思公之惠,立庙以报之。民有入市买菜者,公召谕之曰:"邑居之民,无地种植,且有他业,买菜可也。汝村民,皆有土田,何不自种而费钱买

菜?”笞而遣之。自后人家皆置圃,至今谓芦菔为“张知县菜”。(《梦溪笔谈》)

7)故当时之务,不兴土功,不作师徒,庶人不冠弁、娶妻、嫁女、享祀,不酒醴聚众。农不上闻,不敢私籍于庸,为害于时也。然后制野禁,苟非同姓,农不出御,女不外嫁,以安农也。(《吕氏春秋》)

8)道曰:均地分力,使民知时也。民乃知时日之蚤晏,日月之不足,饥寒之至于身也。是故夜寝蚤起,父子兄弟不忘其功,为而不倦,民不惮劳苦。故不均之为恶也,地利不可竭,民力不可殚。不告之以时而民不知,不道之以事而民不为。与之分货,则民知得正矣。审其分,则民尽力矣。是故不使而父子兄弟不忘其功。(《管子》)

2.将以下两段话译成英文。

1)《山海经》,古书名,作者、成书年代不详。目前看到的《山海经》是由郭璞编订的十八篇本,包括山经5篇(南、西、北、东、中)、海经9篇(海外南、西、北、东4篇和海内经5篇),大荒经4篇(东、南、西、北)。《山海经》记载了许多古代神话传说,虽多怪诞之说,但保留了各地有关地理、历史、民族、医药、巫术、动物、植物、矿产等方面的资料,实为研究上古社会的重要文献。

2)“五行”有三种不同的含义:其一,指五种最基本的事物或构成万物的五种元素。《尚书·洪范》最早明确了“五行”的内容,即金、木、水、火、土。五种事物或元素有其各自的属性,彼此间存在相生相克的关系。其二,五行进一步被抽象为理解万物和世界的基本框架,万物都可以纳入到五行的范畴之中,并因此被赋予不同的性质。其三,指五种道德行为。荀子(前313?—前238)曾指责子思(前483—前402)、孟子(前372?—前289)“按往旧造说,谓之五行”,从郭店楚墓竹简及马王堆汉墓帛书相关文字内容来看,该“五行”指仁、义、礼、智、圣。

3. 以下三段分别记述了栽种桃、梨、栗的育苗方法，根据释文进行翻译。

熟时，合肉全埋粪地中；直置凡地，则不生，生亦不茂。桃性早实，三岁便结子，故不求栽也。至春，既生，移栽实地。若仍处粪中，则实小而味苦矣。

梨熟时，全埋之。经年。至春，地释，分栽之；多著熟粪及水。至冬，叶落，附地刈杀之，以炭火烧头。二年即结子。若穞生及种而不栽者，则著子迟。

栗初熟，出壳，即于屋里埋著湿土中。埋必须深，勿令冻彻。若路远者，以韦囊盛之。停二日已上，及见风日者，则不复生矣。至春二月，悉芽生，出而种之。

（《齐民要术》）

【释文】

桃子成熟时，连皮带肉和核，一齐埋在有粪的地里；就这么种在一般地里，多半不发芽；发芽，苗也不茂盛。桃树结实很早，三岁的树，就可以结果，所以不用扦插。到明年发芽后，移栽到实地里。如果再留在粪地里，果实小，味道也苦。

梨熟了的时候，整个地埋下。让新苗过一年。到春天地解冻后，分开来栽；多用些熟粪作基肥，多浇些水。冬天落叶后，平地面割掉，用炭火烧灼伤口。再过两年，就结实了。野生的树苗和定植未移的实生苗，结实都很迟。

栗刚刚成熟，从壳中剥出后，立即放在房屋里面用湿土埋着。必须埋得够深，不要让它冻透。如果从遥远的地方取种的，用熟皮口袋盛着。凡剥出后在大气中停留过两天以上，见过风和太阳的，都不能发芽。到第二年春天二月间，都已经生了芽，就拿出来种上。

延伸阅读

《吕氏春秋》译文赏析

《吕氏春秋》共有三个英译本,一个是诺布诺克和里格尔的合译本(以下简称"合译"),一个是翟江月译本(以下简称"翟译"),一个是汤博文译本(以下简称"汤译")。

下文选自《吕氏春秋·有始览第一·应同》,记述了五行与农事的关系。

代火者必将水,天且先见水气胜,水气胜,故其色尚黑,其事则水。水气至而不知数备,将徙于土。天为者时,而不助农于下。

合译:The successor to Fire is certain to be Water. Heaven has again first given signs that the ethers of Water are in ascendance. Since the ethers of Water are ascendant, the ruler should honor the color black and model his affairs on Water. If the ethers of Water culminate and no one grasps that fact, the period when it is effective will come to an end, and the cycle will shift to Earth. "Heaven makes the season, but will not assist farmers here below."

翟译:The element that replaces the Element of Fire must be the Element of Water. Omens of water at first must show up out of the will of Heaven. When the Element of Water was prevailing, people of the new dynasty were to respect the color black, and everything was to be done in accordance with the Element of Water. If people did not realize that the Element of Water was prevailing, they could not know that they should all act according to the Element of Water, and thus, it would turn to the Element of Earth. The Five Elements appear by turns according to the will of Heaven. However, they will not change their order to facilitate farming affairs.

汤译:As it was water that would replace fire, heaven would then show signs of water in the ascendant. As the vapor of water is in the ascendant, the black color was to be favored and things were to be done in accordance with the principles of water. When the vapor of water arrives and people do not know that conditions are ready, it will shift to earth. Heaven arranges the four seasons not for the purpose of helping farming on earth alone.

拓展知识

下文简述了《齐民要术》对中国农学的影响，由石声汉撰写。除文中的繁体字转换为简体字外，其余均为原文摘录。

The Influences of *Ch'i Min Yao Shu* on Agricultural Sciences in China

by Shih Shêng-Han

Chia Ssu-hsieh, as we mentioned at the beginning of the present survey, summarised his book in the following way:

"Beginning with ploughing and cultivation, down to the making of vinegar and meat-paste, any art useful in supporting daily life is jotted down."

His summary is indeed a faithful miniature portrait of his encyclopaedia of agricultural science and technology. Starting from reclamation of land as new plots, for all outdoor work such as cultivation of cereals, culinary vegetables, fruits and tinctorial plants, plantation of quick-growing timber wood, animal husbandry, and going down to those indoor subsidiary occupations such as sericulture, food-preparation, also notes on dyeing, making cosmetics and other crafts, and even potential sources of food material to be used in case of famine, instructions were carefully collected, and any personal experience recorded and commented. The sources of the subject-matter, if we interpret the author's statements in present day parlance, may thus be visualised:

"Excerpts from classics or contemporary writings," i.e. written documents;

"Proverbs and folksongs," i.e. verbal information;

"Informations gathered from experts," i.e. existing knowledge;

"Personal experiences," i.e. new evidence.

In other words, Chia Ssu-hsieh systematically organised in his book most of the agricultural knowledge to which he had access. But by adding the results of experimentation, he transformed it into a series of masterly conclusions, and unintentionally advances his subject to the level of real science. His epoch-making masterpiece therefore not only summarises the traditions in an orderly manner, but establishes an excellent standard or model for the later "agriculturists" and "agricultural treatises." The domain of "agriculture" as defined in the *Ch'i Min Yao Shu*, extends over agronomy, sylviculture, horticulture and zootechny, which it regards as founded upon knowledge of astronomy, meteorology, pedology, botany, zoology, chemistry, physics and other branches of natural sciences. Later Chinese agricultural treatises with so comprehensive a scope are naturally rather few in number. We may count (a) 农桑辑要 the *Nung Sang Chi Yao* (Essentials of Agriculture and Sericulture), compiled by the Ministry of Agriculture in the reign of Kublai Khan, (b) 农书 the *Nung Shu* (Treatise on Agriculture) of 王祯 Wang Chen, (c) 农政全书 the *Nung Cheng Ch'uan Shu*, (Comprehensive Treatise on Agricultural Administration) of 徐光启 Hsü Kuang-ch'i, and (d) 授时通考 the *Shou Shi T'ung K'ao* (Book on the Agricultural Timing), compiled by imperial order ca. 1742 as making up the whole list. Of these 4, the former 2 do not surpass the *Ch'i Min Yao Shu* in bulk; the later 2, containing 1.1 and 0.98 million characters respectively, are both about 10 times as large as the *Ch'i Min Yao Shu*. The *Nung Sang Chi Yao* deals in a more detailed way with sericulture, the *Nung Shu* devotes half of its pages to the description of agricultural implements, a rather novel field about which the *Ch'i Min Yao Shu* gives only scattered details. The *Nung Cheng Ch'uan Shu* allots ample space to two further new themes, —hydraulic engineering and famine-relief. While each of the 4 books has one or more special features of its own, these features are not actually totally new or totally unknown in the *Ch'i Min Yao Shu*. On the other hand, these 4 books have at least one thing in common, namely, that quotations from the *Ch'i Min Yao Shu* always form the central or leading matter in most of the subjects treated. Thus we are quite justified in stating that these books ought to be looked upon as orthodox developments of the *Ch'i Min Yao Shu* with extensions or annexations. Most of the minor agricultural treatises of much less bulk appearing later on also quoted generously from the *Ch'i Min Yao Shu* or simply patterned themselves upon it. It is therefore only fair to say that the *Ch'i Min Yao Shu* is the impetus, initiative and prototype of the whole tradition of extant Chinese agriculturalistic literature.

练　习

1. According to Chia Ssu-hsieh's summary, what aspects of daily life are covered in *Ch'i Min Yao Shu*?

   A. Only outdoor work.

   B. Only indoor work.

   C. Both outdoor and indoor work.

   D. Only animal husbandry.

2. What transformation did Chia Ssu-hsieh make to the agricultural knowledge in *Ch'i Min Yao Shu*?

   A. He advanced it to the level of real science.

   B. He organized it systematically.

C. He included more personal experiences.

D. He focused on animal husbandry.

3. According to the passage, the *Ch'i Min Yao Shu* encompasses the following branches of natural sciences except ________.

A. chemistry B. physics

C. zoology D. physiology

4. What is the main characteristic shared by the later Chinese agricultural treatises mentioned in the passage?

A. They are more detailed than *Ch'i Min Yao Shu*.

B. They focus on agricultural implements.

C. They quote extensively from *Ch'i Min Yao Shu*.

D. They do not mention animal husbandry.

5. How does the author describe the relationship between the later Chinese agricultural treatises and *Ch'i Min Yao Shu*?

A. They are entirely new developments.

B. They are extensions or annexations of *Ch'i Min Yao Shu*.

C. They are independent of *Ch'i Min Yao Shu*.

D. They contradict the principles of *Ch'i Min Yao Shu*.

6. What role does the *Ch'i Min Yao Shu* play in the tradition of Chinese agricultural literature according to the author?

A. It is a minor influence.

B. It is the impetus, initiative, and prototype.

C. It is irrelevant to later developments.

D. It is a controversial figure.

第九章　农业科技典籍修辞的翻译

◎本章学习目标◎

1. 了解农业科技典籍修辞的特点

2. 理解比喻、拟人、押韵、排比的含义和作用

3. 掌握农业科技典籍修辞的翻译方法

农业科技典籍中运用了许多修辞技巧，比如前面章节中的例子“作鱼眼沸汤以淋之”，“鱼眼沸”指水初沸时冒出像鱼眼一样的气泡；再比如“芜菁根其大如碗口”，形象地描述了芜菁根的大小。另外，文言文讲究韵律的美感和句式的齐整，押韵和排比也是常用的修辞技巧。

第一节　比喻的翻译

【例1】

造花盐、印盐法：五六月中，旱时，取水二斗，以盐一斗投水中，令消尽，又以盐投之。水咸极，则盐不复消融。易器淘治沙汰之。澄去垢土，泻清汁于净器中。……好日无风尘时，日中曝令成盐，浮，即接取，便是“花盐”；厚薄光泽似钟乳。久不接取，即成“印盐”，大如豆，正四方，千百相似。成印辄

沉,漉取之。花、印二盐,白如珂雪,其味又美。

Preparation of "snow-flake" and "seal" salt: In the dry hot summer-months (5th and 6th), take 2 斗 tou (≈ 4.4 litre) of water, drop in 1 tou crude salt. Let it dissolve and add some more. Finally, the brine will be so concentrated that no more salt will dissolve. Decant into another vessel, shake and separate the floating debris. Now let it stand until all dirt settles. Decant into a clean container. ...Wait for a windless fine day, (so there will be no flying dust) sun it to obtain pure salt. Pellicles formed on the surface when intercepted is the "snow-flake" salt, resembling stalactite both in lustre and form, and can be removed. If not removed, "seal" salt of the size of beans, perfectly cubic on all sides, will come down in hundreds and thousands of crystals similar to each other. Both snow flake and seal salt are white as snow or opal, and taste excellent.

(《齐民要术》卷八《常满盐、花盐第六十九》,石声汉 译)

原文记述的是古代获取纯净食盐的方法,造"花盐"和"印盐"是食盐的盐底结晶精制法。"淘汰沙汰"意思是将粗盐水中的轻浮灰尘和泥渣等撇掉。"珂"字有两种解释:一种是《玉篇》所说"石次玉也",即白色的燧石、蛋白石、雪花石之类;另一种,是《玉篇》所谓"螺属也",即头足类乃至于腹足类的厚重介壳。总之,都是颜色洁自而具有光泽的不透明固体。

数量词"二斗"的翻译采用了音译+汉字+注。"淘治沙汰"是制盐过程中的术语,此处采用了直译。修辞上,原文运用三个比喻形容盐的外观:"厚薄光泽似钟乳""大如豆""白如珂雪"。喻体"钟乳""豆""珂""雪"在英语中均有对等物,翻译时采用了不同句型进行直译。类似的例子还有:

【例2】

麻子,海东毛罗岛来者,大如莲实,最胜;其次出上郡、北地者,大如豆;南地者子小。

Mazi imported from Maoluo Island to the east of the sea is as big as a lotus seed. It is the best drug. The next is the drug from Shangjun. The drug produced in the north is as big as a soybean. That produced in the south has a smaller seed.

（《本草纲目》，罗希文 译）

【例3】

其树如瓜芦，叶如栀子，花如白蔷薇，实如栟榈，茎如丁香，根如胡桃。

Its trunk is suggestive of the gourd and its leaves, of the gardenia. The flower is like that of the wild red rose turned white. The seeds are like those of the coir palm. The leaves have the fragrance of cloves while the roots are as those of the walnut.

（《茶经·一之源》，Francis Ross Carpenter 译）

原文运用了比喻和排比，用熟悉的植物描述茶树的外观形态，形象生动。译文使用了 be suggestive of、like、have sth. of、as 等多种表示比喻的词汇。

【例4】

一曰拔绒，乃毳毛精细者，以两指甲逐茎挦下，打线织成褐。此褐织成，揩面如丝帛滑腻。

The other kind is called "picked wool". It is the finest and obtained by hand-picking one by one. Cloth made of it feels as fine and smooth as silk.

（《天工开物·乃服第二卷》，王义静、王海燕、刘迎春 译）

原文的意思是：另一种叫拔绒，是细毛中最精细的，用两指甲逐根从羊身上拔下，再打线织成绒毛布。这种毛布织成后，手摸布面就像丝帛那样光滑细腻。

【例5】

凡玉唯白与绿两色。绿者中国名菜玉，其赤玉、黄玉之说，皆奇石、琅玕之类。价即不下于玉，然非玉也。凡玉璞，根系山石流水。未推出位时，璞中玉软如绵絮，推出位时则已硬，入尘见风则愈硬。谓世间琢磨有软玉，则又非也。

Jade has only two colors: white and green. Green jade is called "vegetable jade" in the Central Plains. The so-called red jade and yellow jade are in effect unique stones. Lang Gan (jade-like stones) is as expensive as jade but are not jade at all. The stones containing jade are found in the rushing water coming down from mountains. The jade within these stones is as soft as cotton. As soon as the jade is cut open, it becomes hard and even becomes harder after it is carried to the river mud and fanned by the wind. As a result, it is wrong to say that there is a kind of "soft jade" that can be polished and used.

(《天工开物·珠玉第十八卷》，王义静、王海燕、刘迎春 译)

原文的意思是：玉只有白、绿两种颜色，绿玉在中原地区叫菜玉。所谓赤玉、黄玉之说，都指奇石、琅玕之类，虽然价钱不下于玉，但终究不是玉。含玉之石产于山石流水之中，未剖出时璞中之玉软如绵絮，剖露出来后就已变硬，遇到风尘则变得更硬。世间有所谓琢磨软玉的，这又错了。

【例6】

其沸，如鱼目，微有声，为一沸；缘边如涌泉连珠，为二沸；腾波鼓浪，为三沸。已上水老不可食也。

The so-called simmer of water, or the initial boil, is a state when tiny fisheye-like bubbles appear with a low rustling sound. Seething is the second stage when strings of crystal beads gush out around the edge of the wok. The third and last stage refers to the ebullient state of water marked with lumpy waves. This draws the line. Exceeding this point, the water is assumed overdone, and should no longer be used for tea.

(《茶经·五之煮》，姜怡、姜欣 译)

原文记述了煮茶时对水的沸腾程度的掌握，水加热到冒出鱼眼睛似的水泡，并轻声吟唱时，称作第一沸；当锅边涌出连珠般的气泡时，称作第二沸；当锅中波涛翻滚时，称作第三沸，此时的水要是再继续煮就过头了，不再宜于饮用。

【例7】

故礼之于人也，犹酒之有糵也，君子以厚，小人以薄。

Therefore the rules of propriety are for man what the yeast is for liquor. The superior man by (his use of them) becomes better and greater. The small man by his neglect of them becomes meaner and worse.

（《礼记·礼运》，James Legge 译）

原文的意思是：礼对于人来说，就像酿酒必须有酒曲一样。酒曲厚酒就醇美，酒曲薄酒味也淡薄。君子厚于礼，所以成为君子；小人薄于礼，所以成为小人。

【例8】

六经为川，肠胃为海，九窍为水注之气。以天地为之阴阳，阳之汗，以天地之雨名之；阳之气，以天地之疾风名之。暴气象雷，逆气象阳。故治不法天之纪，不用地之理，则灾害至矣。

The Six-Channels act as mountains and valleys, the intestines and the stomach act as the seas, the nine orifices are the regions where Shuiqi (Water-Qi) infuses. To compare the heavens and the earth to Yin and Yang in the human body, sweating induced by the movement of Yang is just like rain in nature, the movement of Yangqi is just like strong wind in nature, a flare of temper of human beings is just like the rumble of thunder, and the adverse flow of Qi in the human body is just like Yang. So if the practice of cultivating health does not follow the principles of the heavens and the rules of the earth, disasters will be caused.

（《黄帝内经·素问》，李照国 译）

原文将人体内的经脉、脏器、情绪等与山川河流、自然现象等相比,并以此解释疾病产生的原因,充分体现了《黄帝内经》中“天人相应”的理念。

练　习

1. 翻译以下典籍原文,特别注意比喻的翻译。

1)然枣,鸡口,槐、兔目,桑、虾蟆眼,榆、负瘤散,自余杂木,鼠耳、蝱翅,各其时。此等名,即皆是叶生形容之所象似。(《齐民要术》)

2)故人情者,圣王之田也。修礼以耕之,陈义以种之,讲学以耨之,本仁以聚之,播乐以安之。(《礼记》)

3)一种绿豆,圆小如珠。绿豆必小暑方种,未及小暑而种,则其苗蔓延数尺,结荚甚稀。若过期至于处暑,则随时开花结荚,颗粒亦少。(《天工开物》)

4)小平似覆釜,一边光彩微似镀金者,此名珰珠,其值一颗千金矣。(《天工开物》)

5)幼珠如粱粟,常珠如豌豆。埤而碎者曰玑。自夜光至于碎玑,譬均一人身,而王公至于氓隶也。(《天工开物》)

6)后世方土效灵,人工表异,陶成雅器,有素肌、玉骨之象焉。(《天工开物》)

7)茶芽,古人谓之雀舌、麦颗,言其至嫩也。今茶之美者,其质素良而所植之土又美,则新芽一发便长寸余,其细如针,唯芽长为上品,以其质干、土力皆有余故也。如雀舌、麦颗者,极下材耳,乃北人不识,误为品题。予山居有《茶论》,《尝茶》诗云:“谁把嫩香名雀舌,定知北客未曾尝。不知灵草天然异,一夜风吹一寸长。”(《梦溪笔谈》)

8)枸杞,陕西极边生者高丈余,大可作柱,叶长数寸,无刺,根皮如厚朴,甘美异于他处者,《千金翼》云“甘州者为真,叶厚大”者是。大体出河西诸郡,其次江池间圩埂上者,实圆如樱桃,全少核,暴干如饼,极膏润有味。(《梦溪笔谈》)

9)荞麦南北皆有。立秋前后下种，八九月收刈，性最畏霜。苗高一二尺，赤茎绿叶，如乌桕树叶。开小白花，繁密粲粲然。结实累累如羊蹄，实有三棱，老则乌黑色。(《本草纲目》)

10)(饴糖)因色紫类琥珀，方中谓之胶饴，干枯者名饧。(《本草纲目》)

11)沫饽，汤之华也。华之薄者曰沫，厚者曰饽，细轻者曰花。如枣花漂漂然于环池之上，又如回潭曲渚青萍之始生，又如晴天爽朗有浮云鳞然。其沫者，若绿钱浮于水湄，又如菊英堕于尊俎之中。饽者，以滓煮之，及沸，则重华累沫，皤皤然若积雪耳。(《茶经》)

2.将以下两段话译成英文。

1)《山海经》是一部内容丰富的中国远古社会百科全书，内容涉及历史、地理、民族、神话、宗教、动物、植物、矿产、医学等。《山海经》包罗之广、内容之奇诡，历代书籍罕有匹敌。全书共约31 000字，分为十八卷，从卷一“南山经”到卷五“中山经”称为《五藏山经》，从卷六“海外南经”到卷十八“海内经”称为《海经》。两部分合起来，总称《山海经》。《山经》以五方山川为纲，记述的内容包括古史、草木、鸟兽、神话、宗教等。《海经》除著录地理方位外，还记载远国异人的风貌。全书记载了100余个国家、近3 000地名、447座山、300余条水道、204个神话人物、300多种怪兽、400多种植物、100余种金属和矿物。

2)颛顼(公元前2342年—公元前2245年)，出生于若水(今蜀地)，黄帝之孙，昌意之子，是中国上古时代部落联盟首领，“五帝”之一。颛顼辅佐少昊有功，封地高阳(今河南省杞县高阳镇)。少昊死后，颛顼与争夺帝位的共工大战于澶渊(今河南省濮阳西)，共工大败，颛顼统一华夏。之后颛顼将中国划为九州，形成各民族真正统一的局面。颛顼还在天文历法上有重要贡献，完成了《颛顼历》。

3. 以下三段分别记述了新鲜葡萄、梨、栗的保存方法,根据释文进行翻译。

1)藏葡萄法:极熟时,全房折取。于屋下作荫坑,坑内近地,凿壁为孔,插枝于孔中,还筑孔使坚。屋子置土覆之,经冬不异也。

2)藏梨法:初霜后,即收。霜多,则不得经夏也。于屋下掘作深荫坑,底无令润湿;收梨置中,不须覆盖,便得经夏。摘时必令好接,勿令损伤。

3)藏生栗法:著器中。晒细沙可燥,以盆覆之。至后年五月,皆生芽而不虫者也。

(《齐民要术》)

【释文】

1)藏鲜葡萄法:葡萄极熟时,整丛地摘下来。在屋子里面掘一个不见光的坑,在坑四边近底的地方,壁上凿许多小孔,把果丛柄插进去,再用泥筑紧。屋里边坑上堆土,盖着,过一个冬还不会变。

2)藏梨法:初霜过后,赶快收摘。经霜次数多,就不能过夏天了。在屋里掘一个不见光的深坑,坑底要干。梨就放在坑里,不必盖,也可以过明年夏天。摘的时候,一定要好好地接着,让梨掉下时不碰伤,便易保存。

3)保存新鲜栗的方法:放在容器里。加上晒干了的细沙,用瓦盆盖在口上。到第二年五月,都发芽了,但不生虫。

延伸阅读

《山海经》译文赏析

下文选自《山海经·大荒西经》。两个英译本分别为:安妮·比勒尔(Anne Birrell)译本(以下简称"比译");王宏、赵峥合译本(以下简称"合译")。

西北海之外,大荒之隅,有山而不合,名曰不周负子,有两黄兽守之。有水曰寒暑之水。水西有湿山,水东有幕山。有禹攻共工国山。

比译：Beyond the northwest seas, at the corner of the Great Wilderness, there is a mountain that is deformed. Its name is Mount Notround-borechild. There are two yellow beasts guarding it. There is a river here called the River Coldhot. To the west of this river there is Mount Scorched. To the east of the River Coldhot there is Mount Tent. Besides, there is the mountain Where Yü Attacked the Country of Common Work.

合译：Beyond the Northwest Sea and at a corner of the Great Wilderness there is a mountain called *Buzhoufuzi* which is deformed and is guarded by two yellow animals. There is a river called *Hanshu*, which means that its water is half cold and half warm. To its west there is Mount *Shishan* while to its east there is Mount *Mushan*. Besides, there is a mountain called Yu-Attacking-the-Kingdom-of-Gonggong.

有国名曰淑士，颛顼之子。

比译：There is a country here. Its name is Fineknight. The people here are the children of the great god Fond Care.

合译：There is a kingdom called Shushi. Its people are all descendants of King Zhuanxu.

有神十人，名曰女娲之肠，化为神，处栗广之野，横道而处。

比译：There are deities here, ten gods in all. Their name is the Guts of Girl Kua. The guts of the goddess Girl Kua turned into gods and they live in the Wilderness of Fullwide. These ten gods took a crosswise route and settled in that place.

合译：There are ten gods who are collectively called Intestines of Nü Wa as they are transformed from the intestines of *Nü Wa*. They live in a wilderness called *Liguang* and stay crosswise on the road.

有人名曰石夷,来风曰韦,处西北隅以司日月之长短。有五采之鸟,有冠,名曰狂鸟。

比译:There is someone named Stone Pacified. The wind that comes there is called Opposing Wind. Stone Pacified stays at the northwest corner of the world in order to preside over the length and shortness of the journeys of the sun and the moon. There are the birds of five colours in this place. They wear an official cap. Their name is the Mad Birds.

合译:There is a man called Shiyi. The wind that comes is called *wei*. Shiyi lives at the northwest corner, presiding over the distance of the journeys of the sun and the moon. There is a bird of five colors which wears an official cap and is called crazy bird.

有西周之国,姬姓,食谷。有人方耕,名曰叔均。帝俊生后稷,稷降以百谷。稷之弟曰台玺,生叔均。叔均是代其父及稷播百谷,始作耕。

比译:There is the Country of Westround. Its people have Lady as their family name. They eat millet. There is someone just now ploughing. His name is Reap Even. The great god Foremost gave birth to Sovereign Millet. The deity Millet came down on earth bringing the hundred grains. Millet's younger brother is called Round Seal. Round Seal gave birth to Reap Even. Reap Even succeeded his father and Millet. Reap Even sowed the hundred grains and he was the first to invent ploughing.

合译:There is the kingdom of Western Zhou. Its people all take Ji (姬) as their family name and they eat grains. There is a man who is plowing the farmland and is called Shujun. King Jun gave birth to Houji who brought a hundred grains from heaven to the mortal world. Houji's younger brother is Taixi who gave birth to Shujun. Shujun sowed the hundred grains on behalf of his father and Houji and he was the first to engage in plowing.

大荒之中，有山名约大荒之山，日月所入。有人焉三面，是颛顼之子，三面一臂，三面之人不死，是谓大荒之野。

比译：In the middle of the Great Wilderness there is a mountain. Its name is Mount Greatwilds. This is where the sun and the moon set. There are people on this mountain who have three faces. They are the children of the great god Fond Care. They have three faces and a single arm. These three-faced people never die. The place where they are is called the Wilderness of Greatwilds.

合译：Inside the Great Wilderness there is a mountain called *Dahuang*. This is where the sun and the moon set. There are people who have three faces. They are the descendants of King Zhuanxu. They have three faces and one arm, and never die. Their dwelling place is called *Dahuang*.

有互人之国。炎帝之孙名曰灵恝，灵恝生互人，是能上下于天。有鱼偏枯，名曰鱼妇。颛顼死即复苏，风道北来，天乃大水泉，蛇乃化为鱼，是为鱼妇。颛顼死即复苏。

比译：There is the country of the Beside people. The grandson of the great god Flame had the name Divinepower Heartless. Divinepower Heartless gave birth to the Beside people. These people are able to go up to the sky and come back down again. There is a fish here which is withered down one side. Its name is the Fishwife. This is none other than the great god Fond Care. When Fond Care died, he immediately came back to life in this form. Whenever the wind comes from the north, the sky will release a huge flow of water into the springs and then snakes will turn into fish. Such is the Fishwife; she is none other than Fond Care who died and then came back to life in this form.

合译：There is the kingdom of *Huren*. The grandson of Emperor Yandi is called Lingqi. Lingqi gave birth to *Huren* who was able to go up to the sky and came down again. There is a fish which is withered down one side and is called *yufu*. King Zhuanxu revives immediately after his death. When the wind blows from the north, the spring water will pour out from the sky and the snake will turn into fish. This is *yufu*. King Zhuanxu revives immediately after his death.

第二节 拟人的翻译

【例1】

天生五谷以育民，美在其中，有“黄裳”之意焉。稻以糠为甲，麦以麸为衣。粟、粱、黍、稷，毛羽隐焉。

Nature provides five types of grains to nourish people. Grains are hidden in the yellow chaff, and look as beautiful as if they were in yellow robes. Rice is covered in chaff, wheat is enclosed by bran, and millet and sorghum grains are hidden in feather like husks.

(《天工开物·粹精第四卷》，王义静、王海燕、刘迎春 合译)

自然界生长五谷以养育人，而谷粒包藏在黄色谷壳里，像身披“黄裳”一样美。稻以糠为壳，麦以麸为皮。粟、粱、黍、稷的子实都隐藏在毛羽里面。粮食作物的果实像人一样穿上了各种样式的外衣，译文保留了这些生动形象的描写。

【例2】

作蘖法：八月中作。盆中浸小麦，即倾去水，日曝之。一日一度著水，即去之。脚生，布麦于席上，厚二寸许。一日一度，以水浇之。牙生便止。即散收，令干。勿使饼！饼成，则不复任用。此煮白饧蘖；若煮黑饧，即待芽生青成饼，然后以刀㔭取干之。欲令饧如琥珀色者，以大麦为其蘖。

How to sprout wheat: In the 8th month, soak wheat kernel in a pan, drain off the superfluous water, and sun it. Flush with water and drain at once every day. Spread on a mat to the thickness of 2 ts'un (5 or 6 cm.) When the "toes" (i. e. the radicula) appear, sprinkle once a day. After bursting out of the "teeth" (i. e. sprout or coleoptile), collect and dry up. Don't make a "cake" (i. e. interlocking mass) of them! Once entangled, they become useless. These are good for preparation of white barley sugar. To make dark barley sugar, wait till the "teeth" turns green and the corn forming a cake, then cut into small pieces to dry. To prepare amber-colored barley sugar, use barley (instead of wheat).

（《齐民要术》卷八《黄衣、黄蒸及蘖第六十八》，石声汉译）

原文记述的是作蘖米的方法。"蘖"（niè）是麦、豆等的芽，蘖米指生芽的麦粒。使淀粉糖化制糖，必须有淀粉酶的催化。从前靠发芽的谷物种实，即"蘖米"或"蘖"，供给淀粉酶类。"蘖"在中国是很古老的发明。"饧"（xíng）是麦芽或谷芽熬成的饴糖。"㔭"（lí）指用刀分割。

麦粒萌发的过程就像先长脚，后长牙。"脚"指幼根，形状部位都像脚。"牙"指芽鞘，形状颜色都像牙。译文保留了源语中的拟人形象，分别直译为toes和teeth，然后加文内注进行解释。

【例3】

凡苗吐穑之后，暮夜鬼火游烧，此六灾也。此火乃腐木腹中放出。凡木母火子，子藏母腹，母身未坏，子性千秋不灭。

When the rice seedlings are earring up, they will be burned at night by will-o'-the-wisp. This is the sixth disaster. This kind of fire comes from the rotten wood. Fire is dormant in wood and fire exists inside and does not come out when wood is not rotten.

(《天工开物·乃粒第一卷》,王义静、王海燕、刘迎春 合译)

原文记述了水稻生产过程中的灾害之一"鬼火"。这种火是从朽烂的木头中放出的。木生火,火藏于木中,木未坏而火便在其中永不消失。原文用母子形容木火,译文并未保留拟人修辞,而是直接译出木和火的关系。类似的例子还有:

【例4】

故天有精,地有形;天有八纪,地有五里,故能为万物之父母。

So the heavens has Jing (Essence-Qi) and the earth has forms. The heavens demonstrates the eight solar terms and the earth displays the rules of Wuxing (Five-Elements). That is why the heavens and the earth are regarded as the parents source of all things.

(《黄帝内经·素问》,李照国 译)

【例5】

脾主运化,胃司受纳,通主水谷,故皆为仓廪之官。

The spleen governs the transportation and transformation of food and drinks, and the stomach receives them. Both are in charge, so the spleen and the stomach are described as granary officers.①

(《类经·藏象类》)

"仓廪之官"本义是管理粮食仓库的官吏。脾和胃在功能上相互配合,人体摄入的饮食物经过胃的腐熟、消化,再经过脾的运化,转化为精气、血、津液输送到全身,为人体提供营养,脾和胃正是扮演了粮仓管理者的角色。

① 译文来源:中华思想文化术语库"仓廪之官"词条。

练　习

1. 翻译以下典籍原文，特别注意拟人的翻译。

1)麻勃如灰，便收。未勃者收，皮不成；放勃不收，而即骊。(《齐民要术》)

2)大豆戴甲而生，不用深耕。……厚则折项，不能上达，屈于土中而死。(《氾胜之书》)

3)每逢多雨之年，孤野墓坟多被狐狸穿塌。其中棺板为水浸，朽烂之极，所谓母质坏也。火子无附，脱母飞扬。(《天工开物》)

4)脾……仓廪之本，营之居也……其华在唇四白，其充在肌。(《黄帝内经》)

5)其始，若茶之至嫩者，蒸罢热捣，叶烂而牙笋存焉。假以力者，持千钧杵亦不之烂。如漆科珠，壮士接之，不能驻其指。及就，则似无穰骨也。炙之，则其节若倪倪如婴儿之臂耳。(《茶经》)

2. 将以下两段话译成英文。

1)《本草纲目》由明代李时珍所著，是中药学集大成之作，系统总结了明代以前的药物学成就，其中收录药物种类丰富，涉及草部、木部、果部等，广泛用于外感内伤诸病及预防保健。《本草纲目》在中药学发展史上具有划时代意义，该书不仅是中医药学者格物明理，知行合一理学精神的体现，也对中医学理论发展与中医药疗法拓展做出了重要贡献。

2)《黄帝内经》，即《素问》与《灵枢》之合称，是中国现存最早的医学典籍，反映了中国古代的医学成就，奠定了中国医学发展的基础，成为中国医药之祖、医家之宗。所以唐人王冰在《黄帝内经素问注·序》中说："其文简，其意博，其理奥，其趣深；天地之象分，阴阳之候列，变化之由表，死生之兆彰；不谋而遐迹自同，勿约而幽明斯契；稽其言有征，验之事不忒。"在中国几千年漫长的历史中，《黄帝内经》一直指导着中国医学的发展，中医学中众多流派的理论观点,莫不源于《黄帝内经》的基本思想。

3. 下文记述了茶的饮用品赏,根据释文进行翻译。

於戏!天育万物,皆有至妙,人之所工,但猎浅易。所庇者屋,屋精极;所著者衣,衣精极;所饱者饮食,食与酒皆精极之。凡茶有九难:一曰造,二曰别,三曰器,四曰火,五曰水,六曰炙,七曰末,八曰煮,九曰饮。阴采夜焙,非造也;嚼味嗅香,非别也;膻鼎腥瓯,非器也;膏薪庖炭,非火也;飞湍壅潦,非水也;外熟内生,非炙也;碧粉缥尘,非末也;操艰搅遽,非煮也;夏兴冬废,非饮也。

《茶经·六之饮》

【释文】

是啊,大自然孕育了世间万物,林林总总皆有其精妙之处,只是人类所擅长的仅仅是肤浅而有限的几个方面。居住要盖房修屋,人类的建筑已能造得富丽堂皇;驱寒要穿衣戴帽,人类的服饰已能缝得精美绝伦;充饥要开伙造饭,人类的烹饪酿酒之道也考究到了极致。而对有关茶的诸项事宜,人们却还需要摸索出九道难以掌控的关口:一是采制,二是鉴别,三是器具,四是用火,五是择水,六是炙烤,七是碾末,八是烹煮,九是品饮。阴天采摘、夜间焙制,称不上是真正的茶采制;口嚼尝味、鼻闻辨香,称不上是真正的茶鉴别;锅沾荤腥、碗存异味,称不上是真正的茶器具;油质树枝、厨存余炭,称不上是真正的茶柴火;湍流截取、淤潭汲舀,称不上是真正的茶用水;外熟里生、外焦里硬,称不上是真正的茶炙烤;舂压过细、末如青粉,称不上是真正的茶碾磨;操具笨拙、翻搅慌促,称不上是真正的茶烹煮;夏天常用、冬季摒弃,称不上是真正的茶品饮。

延伸阅读

《茶经》译文赏析

下文选自《茶经·三之造》。三个英译本分别为：弗朗西斯·罗斯·卡彭特(Francis Ross Carpenter)译 *The Classic of Tea*(以下简称"卡译")；姜怡、姜欣合译 *The Classic of Tea*(以下简称"姜译")；邱贵溪译 *The Classic on Tea*(以下简称"邱译")。

茶有千万状，卤莽而言，如胡人靴者，蹙缩然[京锥文也]；犎牛臆者，廉襜然；浮云出山者，轮囷然；轻飙拂水者，涵澹然；有如陶家之子，罗膏土以水澄泚之[谓澄泥也]；又如新治地者，遇暴雨流潦之所经。此皆茶之精腴。有如竹箨者，枝干坚实，艰于蒸捣，故其形簏簁然[上离下师]；有如霜荷者，茎叶凋沮，易其状貌，故厥状委悴然。此皆茶之瘠老者也。

卡译：Tea has a myriad of shapes. If I may speak vulgarly and rashly, tea may shrink and crinkle like a Mongol's boots. Or it may look like the dewlap of a wild ox, some sharp, some curling as the eaves of a house. It can look like a mushroom in whirling flight just as clouds do when they float out from behind a mountain peak. Its leaves can swell and leap as if they were being lightly tossed on wind-disturbed water. Others will look like clay, soft and malleable, prepared for the hand of the potter and will be as clear and pure as if filtered through wood. Still others will twist and turn like the rivulets carved out by a violent rain in newly tilled fields. Those are the very finest of teas. But there are also teas like the husk of bamboo, hard of stem and too firm to steam or beat. They assume the shape of a sieve. Then there are those that are like the lotus after frost. Their stem and leaves become sere and limp, their appearance so altered that they look like piled-up rubble. Such teas are old and barren of worth.

姜译：The surface of caked tea could take on thousands of different looks. Here is an inkling of their appearances: Some crease like the Tartars' leathern boots, others curl like buffalo's dewlap. Some unfold like a cluster of floating clouds from behind mountains, while others ripple almost audibly like a river being fondled by a breeze. Some look sleek and silky like pottery clay finely sifted and pasted with water, yet others feel rugged and rough like newly cultivated field eroded by pouring rains. All these are good teas in most cases. Some tea leaves are tough as bamboo sheaths with stiff stalks, making it hard to get them thoroughly steamed and finely pounded. The exterior of tea-cakes made of such leaves may often end up poriferous and netty like a coarse sieve. Some tea leaves are withered and blighted like frost-bitten lotus. Tea-cakes made from such baggy leaves would be sapless and shabby. Obviously, they fall into the category of inferior tea.

邱译：Tea cakes vary in their shapes. Roughly speaking, some are like the boots worn by the Tartars (refer to ethnic minorities inhabiting the northern and western part of China). The crumpled surface is like the venison of the arrowhead. Some are like the chest of buffalo with fluctuating wrinkles; some are like the floating clouds drifting and circling around the mountains; some are like the breeze blowing the water, rippling gently; some are like the fine and smooth clay pastes, sieved and washed by the potter, and some are like the newly levelled ground which is so smooth as if washed by the rainstorm. These are fine and top-quality tea. Some tea leaves are like the crusts of the bamboo shoots, with hard stems difficult to steam and pound. The tea cake made of such tea leaves is like bamboo sieve with hard and rough surface. Some tea leaves are like the withered lotus leaves damaged by frost with the wilted and deformed stalks. The tea cake made of such tea leaves has a withered shape. These are coarse and inferior tea.

自采至于封七经目，自胡靴至于箱荷八等。或以光黑平正言嘉者，斯鉴之下也；以皱黄坳垤言佳者，鉴之次也；若皆言嘉及皆言不嘉者，鉴之上也。何者？出膏者光，含膏者皱；宿制者则黑，日成者则黄；蒸压则平正，纵之则坳垤。此茶与草木叶一也。茶之否臧，存于口诀。

卡译：From picking to sealing there are seven steps, and there are eight categories of shapes, from the leaves that look like a Mongol's boots to those that are like a lotus flower killed by frost. Among would-be connoisseurs there are those who praise the excellence of a tea by noting its smoothness and commenting upon the glossy jet shades of the liquor. They are the least capable of judges. Others will tell you it is good because it is yellow, wrinkled and has depressions and mounds. They are better judges. But the really superior taster will judge tea in all its characteristics and comment upon both the good and the bad. For every individual criticism there is a reason. If the tea leaf exudes its natural juices, it will be glossy and black. If the oils are contained, then it will appear wrinkled. If it has been manufactured for a long time, it will be black. Tea over which the sun has scarcely set will be yellow. Steamed and tamped, it will be smooth. Allowed to remain loose, it will have hollows and hills. There is nothing unnatural in that, for tea is like other herbs and leaves in that regard. Its goodness is a decision for the mouth to make.

姜译：From picking to packing, the tea processing goes through seven procedures. From the "Tartars' boots" to the "frost-bitten lotus," processed tea-cakes are sorted into eight quality grades. Judging by their looks, good comments tend to be given to those with smooth edge and in glossy black tint. However, such hasty judgment is often more prejudiced than precise. Assuming a tea to be good by its wrinkled or rugged yellowish appearance is no more well-grounded. Justifiable opinions should offer the why in addition to the what. Teas with the above-mentioned attributes could be either good or bad. Here comes the explanation for why it is thus said. Tea with its juice pressed out to the surface will surely attain a glossy tint, while those without oozed juice will look dull and rugged. Tea plucked and processed with a night in between would retain the nocturnal shade on the cakes, while intraday treatment will endow tea-cakes with a bright yellow. Tea-cakes tightly pressed in a gui after being steamed soft will be edged smoothly, while those slightly molded will look shapeless. In this aspect, tea is no different from any other foliaceous plants. Whatsoever, the final assessment on the quality of any tea comes from actual sampling.

邱译：There are seven process procedures from plucking the tea leaves to packing the tea cakes and eight grades from the crumped shape of boots-like tea cakes to the withered shape of frost-damaged lotus leaf. It is the inferior discrimination in tea if someone takes the tea cake with glossy black smooth shape as the good one; it is the secondary discrimination in tea if judging by the characteristics of crumpled shape, yellow color and even or uneven surface of the tea cake and etc.; and it is the best discrimination in tea if we can tell the merits and demerits of the tea cake on the whole. Why? The tea cakes look shining if the tea juice is squeezed out and looks crumpled because of the tea juice; the tea cake looks black if it is processed overnight, looks yellow if it is processed on that very day, looks smooth if pressed tightly after being steamed and looks uneven if pressed arbitrarily. The leaves of tea and other plants share the same characteristics. There exists the pithy mnemonic formula for discrimination the good tea from the bad one.

第三节　押韵的翻译

【例1】

丰年多黍多稌。亦有高廪，万亿及秭。为酒为醴，烝畀祖妣，以洽百礼，降福孔皆。

Millet and rice abound this year;

High granaries stand far and near.

There are millions of measures fine;

We make from them spirits and wine

And offer them to ancestors dear.

Then we perform all kinds of rite

And call down blessings from Heaven bright.

（《诗经·周颂·臣工之什·丰年》，许渊冲 译）

丰收年多黍多稻。也有仓库很是高，积粮万万及亿亿。做酒做甜酒都好，进献先祖和先妣，用来配合百礼好，降下福禄都是好。原文是庆丰年、祭田神的乐诗，其中有“秭”、“醴”、“妣”、“礼”押韵，译文也使用了三种不同韵脚，一是 year，near 和 dear，二是 fine 和 wine，三是 rite 和 bright，保留了诗歌的音乐美，读起来朗朗上口。类似的例子还有：

【例2】

大田多稼，既种既戒，既备乃事。以我覃耜，俶载南亩，播厥百谷，既庭且硕，曾孙是若。

Busy with peasants' cares,

Seed selected, tools repaired,

We take our sharp plough-shares

When all is well prepared.

We begin from south field

And sow grain far and wide.

Gross and high grows our yield;

Our lord's grandson's satisfied.

(《诗经·小雅·北山之什·大田》,许渊冲 译)

【例3】

苗:其弱也,欲孤;其长也,欲相与俱;其熟也,欲相扶。是故三以为族,乃多粟。吾苗有行,故速长;弱不相害,故速大;横行必得,从行必术,正其行,通其风。

Young plants at the start

Must first be wide apart.

But when they are half-grown

Need no more stand alone,

And when they are ripe and old

Can one another uphold.

Bunches in threes make no harm,

They bring good harvest on the farm.

Our plants being set in rows,

Each of them rapidly grows.

As none does its neighbours' oppose,

In size the young plants rapidly grow.

Upstanding in straight street,

The good fresh air they meet.

（《吕氏春秋》，石声汉 译）

有时原文中并没有明显的押韵，在翻译时为了体现诗歌的文体特征，译文可以采用押韵修辞。比如以下两例：

【例4】

左思《娇女》诗："吾家有娇女，皎皎颇白皙；小字为纨素，口齿自清历。有姊字惠芳，眉目粲如画；驰骛翔园林，果下皆生摘。贪华风雨中，倏忽数百适；心为茶荈剧，吹嘘对鼎䥶。"

The civil official Zuo Si in the Jin Dynasty once wrote a poem *My Cute Girls*, of which a few lines are quoted as follows:

My two daughters are cute girls,
Fair and flawless as lily pearls.
We give the younger the name Pure,
Her tongue's glib but never demure.
The elder's name is an orchid fine,
Brows are rainbow and eyes shine.
They brisk in woods like two fairies,
Can't wait to get ripe fruits and berries.
To flowery nature they're so much bound,
Wind and rain chorus a cheerful sound.
Tea scents from home lure them with desire,
Pursing rosy lips they help blow the fire.

（《茶经·七之事》，姜怡、姜欣 译）

【例5】

王荆公尝有诗云：神震冽冰霜，高穴雪与平。空山渟千秋，不出呜咽声。

山风吹更寒,山月相与清。北客不到此,如何洗烦酲。

Wang Anshi has composed an ode on it after he sipped the water:

Chilly fresh the water outshines frost,
A snow lace covers the cave glossed.
Brewed for ages in the deep mountain,
The spring tunes in its lonely fountain.
Iced crystal with each wintry breeze,
Looking purer in moonlight's caress.
Boiling tea not by this blessed wonder,
How could anyone be high and sober?

(《续茶经·茶之煮》,姜怡、姜欣 译)

练 习

1.翻译以下典籍原文,特别注意押韵的翻译。

1)思文后稷,克配彼天。立我烝民,莫匪尔极。贻我来牟,帝命率育。无此疆尔界,陈常于时夏。(《诗经》)

2)嗟嗟臣工!敬尔在公。王釐尔成,来咨来茹。嗟嗟保介!维莫之春,亦又何求?如何新畬?於皇来牟?将受厥明。明昭上帝,迄用康年。命我众人:庤乃钱镈,奄观铚艾。(《诗经》)

3)噫嘻成王,既昭假尔。率时农夫,播厥百谷。骏发尔私,终三十里。亦服尔耕,十千维耦。(《诗经》)

4)张孟阳《登成都楼》诗云:"借问扬子舍,想见长卿庐;程卓累千金,骄侈拟五侯。门有连骑客,翠带腰吴钩;鼎食随时进,百和妙且殊。披林采秋橘,临江钓春鱼;黑子过龙醢,果馔逾蟹蝑。芳茶冠六清,溢味播九区;人生苟安乐,兹土聊可娱。"(《茶经》)

5)松风桂雨到来初,急引铜瓶离竹炉。待得声闻俱寂后,一瓯春雪胜醍醐。(《续茶经》)

6)林逋《烹北苑茶有怀》:石碾轻飞瑟瑟尘,乳花烹出建溪春。人间绝品应难识,闲对《茶经》忆故人。(《续茶经》)

2. 将以下两段话译成英文。

1)《诗经》是中国最早的诗歌总集。共三百零五篇,又称《三百篇》。先秦时称《诗》。汉代以后,儒家列为经典之一,故称《诗经》。约成书于春秋时代,分为"风""雅""颂"三大类。一般认为是西周初年至春秋中叶的作品。诗篇以四言为主,多采用重言、双声、叠韵和赋、比、兴的写作手法,语言优美,描写生动,富有强烈的现实主义创作精神。《诗经》不仅对古代中国文学发展产生巨大影响,而且是研究先秦社会和思想的宝贵资料。

2)"后稷"是西周时期的农神代表,是周人对其文化祖先的一种尊称,寄托着他们对其群族的深刻体会和认知。周人赋予后稷诸多宗教功能,用来证明王权之神圣性及统治的合法性。

3. 以下一首诗选自《诗经》,是周王祭祖祀神的乐歌,根据释文进行翻译。

楚楚者茨,言抽其棘。自昔何为?我蓺黍稷。我黍与与,我稷翼翼。我仓既盈,我庾维亿。以为酒食,以享以祀,以妥以侑,以介景福。

济济跄跄,絜尔牛羊,以往烝尝。或剥或亨,或肆或将。祝祭于祊,祀事孔明。先祖是皇,神保是飨。孝孙有庆,报以介福,万寿无疆!

执爨踖踖,为俎孔硕,或燔或炙,君妇莫莫。为豆孔庶,为宾为客,献酬交错。礼仪卒度,笑语卒获。神保是格,报以介福,万寿攸酢!

我孔熯矣,式礼莫衍。工祝致告,徂赉孝孙。苾芬孝祀,神嗜饮食。卜尔百福,如几如式。既齐既稷,既匡既敕。永锡尔极,时万时亿!

礼仪既备,钟鼓既戒,孝孙徂位,工祝致告。"神具醉止",皇尸载起。鼓钟送尸,神保聿归。诸宰君妇,废彻不迟。诸父兄弟,备言燕私。

乐具入奏,以绥后禄。尔肴既将,莫怨具庆。既醉既饱,小大稽首。神嗜饮食,使君寿考。孔惠孔时,维其尽之。子子孙孙,勿替引之!

(《诗经·小雅·北山之什·楚茨》)

【释文】

植物丛生是蒺藜,那时除刺靠用犁。自古以来做什么?我自种下黍和稷。我的黍子很茂盛,我的稷子很茂密。我的仓库既装满,我的露仓数有亿。用来做酒和吃食,用来供神和祭祀,用来安坐饮酒足嗜,用来助我得大福祉。

众人奔走有节度,祭神洁净你牛羊,用作秋祭及冬祭。有的剥皮有的煮汤,有的陈设有的供场。司仪先祭庙门旁,祭祀的事很洁净。先祖神道最堂皇,作尸的人得安享。孝孙得会有赐赏,报祭用来赐大福,赐的是万寿无疆!

庖人烧火很恭谨,作为器具用大好,有的烧烤有的炒,主妇安静态度好。食器陈列得很多,作宾作客真不少,献酒酬酒相交错。礼节合法极周到,笑着说话都恰好。作尸的人是来了,报祭用来赐大福,用万寿来做答报。

我是很恭敬了,用礼没有过错好。司仪向神来报告,神往赐福孝孙好。馨香祭祀用得到,神爱酒食吃得了。赐你百种幸福好,福来有期又有程。既是整齐又快好,既是正规又坚妙。永远赐你福气好,是万是亿都得到!

礼仪既经完备,钟鼓既经备好。孝孙既已到位,司仪向神祷告。“神都吃醉了”,做尸的人起来了。打鼓敲钟送尸了,做尸的人回去了。诸个宰夫和主妇,撤掉祭神酒席不迟了。诸父兄弟另设席,完备地饮宴私自好。

乐器具备入奏好,用来安享祭后肴。你的肴既已摆好,没有怨言全说好。既喝醉又吃饱,小子大人叩头报道。神爱好饮酒吃肉,使你能够得寿考。很顺礼很及时,你尽礼又尽孝。你的子子孙孙,不要改变长存好。

延伸阅读

《诗经》译文赏析

下文选自《诗经·小雅·北山之什·大田》。三个英译本分别为：威廉·詹宁斯（William Jennings）译 *The Shi King*（以下简称"詹译"）；埃兹拉·庞德（Ezra Pound）译 *The Classic Anthology Defined by Confucius*（以下简称"庞译"）；许渊冲译 *Book of Poetry*（以下简称"许译"）。

大田多稼，既种既戒，既备乃事。以我覃耜，俶载南亩，播厥百谷，既庭且硕，曾孙是若。

既方既皂，既坚既好，不稂不莠。去其螟螣，及其蟊贼。无害我田稚！田祖有神，秉畀炎火。

有渰萋萋，兴雨祁祁。雨我公田，遂及我私。彼有不获稚，此有不敛穧。彼有遗秉，此有滞穗，伊寡妇之利！

曾孙来止，以其妇子，馌彼南亩，田畯至喜。来方禋祀，以其骍黑。与其黍稷，以享以祀，以介景福。

詹译：

Large are the fields, and much there is to sow,
The seed is chosen, all is done with care,
And all being ready, to the work we go;
And here beginning, each with sharpened share,
We break into the southward-sloping land,
And scatter there of every sort of grain;
Anon erect and stately shall it stand,
And thus the Latest Heir his wish obtain.

Anon the ear, and then the full soft seed,
Anon more firm, and fine as it is firm, —
No darnel shall be there, nor noxious weed;
The caterpillar and the cankerworm
And grub and weevil shall be cleared away:
To the young crop shall none bring damage dire.
O ghostly Father of the Fields, we pray,
Take them, and give them to the flames of fire.

The clouds are gathering now, an inky pall,
The rains begin, in mild and gentle showers;
First on the public fields then let them fall,
And after that descend on these of ours!
And yonder will be young ungathered grain,
And here be sheaves we trouble not to bind,
And yonder handfuls suffered to remain,
And here the straggling heads we leave behind; —
These the lone widows for their portion gain.

And now he comes, the Latest of his line,
Whileas the wives and children fetch the food, —
Here where the acres to the south incline.
The Steward, too, arrives, in cheery mood.
He* comes, and the pure sacrifice sets forth
Of victims red and black, and gifts of grain,

To Spirits of the air—of South and North;**
And by these gifts and offerings he shall gain
To blessings great still more of greater worth.

* The "he" refers to the personage of the first line.

** The red bull was offered to the Spirits of the South; the black one to those of the North.

庞译：

Great plowlands need
many chores, seed,
tools and forecare.
Grind share and go
start with the plow on south slopes now.
Let the grain grow
then pile it high in courtyard where
As grandsire was, is now heir.

Come sprout, come ear,
hard grain and good
let every weed and tare,
gnaw-bug and worm,
caterpillar, slug
fall dead in flame,
honour to Tien Tsu,
in fact and name,

God of the field.

Thickens the cloudy sky
that rain like a slanting axe
feed our Duke's field
then bless our yield,
We reap not miserly,
old women and poor follow our spoor.
To their relief
leave loosened sheaf,
short stock and unripe ear.

Now's come Greatgrandsire's heir,
women and youngsters bring
lunches to the men labouring
on the south slope, the overseer
does reverence to the four Corners of Air;
pours back the wine to earth (that gave the wine)
red bullock and black
pay for the millet crop;
by offer and sacrifice
funnel us further felicities.

许译:

*Busy with peasants' cares,
Seed selected, tools repaired,

We take our sharp plough-shares
When all is well prepared.
We begin from south field
And sow grain far and wide.
Gross and high grows our yield;
Our lord's grandson's satisfied.

The grain's soft in the ear
And then grows hard and good.
Let nor grass nor weed appear;
Let no insects eat it as food.
All vermins must expire
Lest they should do much harm.
Pray gods to put them in fire
To preserve our good farm,

Clouds gather in the sky;
Rain on public fields come down,
It drizzles from on high
On private fields of our own.
There are unreaped young grain
And some ungathered sheaves,
Handfuls left on the plain
And ears a widow perceives.
And gleans and makes a gain.

Our lord's grandson comes here.
Our wives bring food to acres south
Together with their children dear;
The overseer opens mouth.
We offer sacrifice
With victims black and red,
With millet and with rice.
We pray to fathers dead
That we may be blessed thrice.

* The first stanza described the farm work in spring, the second that in summer, the third harvest in autumn and the last sacrifice in winter.

第四节　排比的翻译

【例1】

故先王之制:四海云至,而修封疆;虾蟆鸣、燕降,而通路除道矣;阴降百泉,则修桥梁。

Therefore, the ancient sagacious kings laid down that: when clouds rise up from the horizon, fortifications on the frontiers should be seen to; when frogs croak and swifts descend, thoroughfares should be cleared; when waters are low, bridges should be repaired.

(《淮南子》,石声汉 译)

原文用排比句式列举了不同时节组织人民从事的不同劳作,译文用

when..., ... should be...保留了原文的排比。

【例2】

国君春田不围泽;大夫不掩群,士不取麛卵。

The ruler of a state, in the spring hunting, will not surround a marshy thicket, nor will Great officers try to surprise a whole herd, nor will other officers take young animals or eggs.

(《礼记·礼运》,James Legge 译)

这一句体现了中国古代生态保护的思想。国君在春天举行田猎,不可合围猎场;大夫不杀尽群居的野兽,士不获取幼兽和鸟卵。原文在句式上是三个否定句的排比,译文使用了not... nor... nor的句型。

【例3】

错国于不倾之地,积于不涸之仓,藏于不竭之府,下令如流水之原。使民于不争之官,明必死之路,开必得之门。不为不可成,不求不可得,不处不可久,不行不可复。

Establish the state on a very stable foundation. Store the food supply in inexhaustible granaries. Lay up the treasure in unlimited depots. Make sure that orders issued by the sovereign are like water flowing from a fountain (the orders will be carried out just like the water will get its supply from the fountain forever). Appoint people to indisputable positions. Clarify the death penalty. Open the door for awards when contributions are accomplished. Do not take action that will never be achieved. Do not seek things that are unreachable at all. Do not stay at places that are unsafe. Do not take action that is not allowed to be repeated again.

(《管子·牧民第一》,翟江月 译)

【例4】

翼而飞,毛而走,呿而言,此三者俱生于天地间,饮啄以活,饮之时义远矣哉! 至若救渴,饮之以浆;蠲忧忿,饮之以酒;荡昏寐,饮之以茶。

Fledged birds are able to soar; fur-bearing beasts are able to scurry; and language-bestowed humans are able to speak. These three dominant beings on earth have long been in existence, eating and drinking to survive. So the vital role of beverage is self-evident. To quench thirst and dryness, water is imbibed; to relieve sorrow and annoyance, wine is guzzled; and to get over fatigue and drowsiness, tea is sipped.

(《茶经·六之饮》,姜怡、姜欣 译)

【例5】

唐陆羽《六羡歌》:不羡黄金罍,不羡白玉杯,不羡朝入省,不羡暮入台。千羡万羡西江水,曾向竟陵城下来。

Lu Yu in the Tang Dynasty wrote in "The Six Envies":

I envy not gold pots bright,

I envy not jade cups white,

I envy not court goers for morning rite,

I envy not officials off work in twilight.

But how, how I envy the West River tide

Flowing via Jingling, my dear home site.

(《续茶经·茶之煮》,姜怡、姜欣 译)

【例6】

盖自古中国所以为衣者,丝麻葛褐,四者而已。汉唐之世,远夷虽以木棉入贡,中国未有其种,民未以为服,官未以为调。宋元之间,始传其种入中国。

For previous to that time, the Chinese used only Silk, Hemp, Flax, and plaited Hemp, to make garments. In the time of the Hán and T'áng dynasties, distant foreigners brought tribute of cotton to China, but it was not yet planted in the empire, nor did the people make clothes of it, neither had the authorities levied taxes on it; and only during the Sung and Yuen dynasties was it first introduced and planted.

（《农政全书》卷三十五《蚕桑广类·木棉》, C. Shaw 译）

【例7】

精神不进，志意不治，故病不可愈。今精坏神去，荣卫不可复收。何者？嗜欲无穷，而忧患不止，精气弛坏，荣泣卫除，故神去之而病不愈也。

Declination of Jingshen (Essence and Spirit) and distraction of Yizhi (mind) make it difficult to treat diseases. Now Jing (Essence) is damaged, Shen (Spirit) is lost and Rongwei (Nutrient-Qi and Defensive-Qi) is out of control. What is the reason? This is exclusively caused by insatiable avarice and excessive anxiety that leads to decay of the Jingshen (Essence and Spirit), scantiness of Rong (Nutrient-Qi) and dysfunction of Wei (Defensive-Qi). That is why Shen (Spirit) is lost but the disease is not cured.

（《黄帝内经·素问》，李照国 译）

原文中并没有典型的排比句式，但译文采用了工整的排比，将“精”“神”“荣”“卫”单独成句，体现了四者在人体健康中的不同表现。

练 习

1. 翻译以下典籍原文，特别注意排比的翻译。

1）岁凶，年谷不登，君膳不祭肺，马不食谷，驰道不除，祭事不县。大夫不食粱，士饮酒不乐。（《礼记》）

2）故治国不以礼，犹无耜而耕也；为礼不本于义，犹耕而弗种也；为义而不讲之以学,犹种而弗耨也;讲之于学而不合之以仁,犹耨而弗获也;合之以仁而不安之以乐,犹获而弗食也;安之以乐而不达于顺,犹食而弗肥也。(《礼记》)

3)强本而节用,则天不能贫;养备而动时,则天不能病;修道而不贰,则天不能祸。(《荀子》)

4)采茶欲晴,藏茶欲燥,烹茶欲洁。(《续茶经》)

5)天下有好茶,为凡手焙坏。有好山水,为俗子妆点坏。有好子弟,为庸师教坏。真无可奈何耳。(《续茶经》)

6)夫以一邑渐及之他邑,何难? 既能其一,进之其十,何难? 由下品而中,由中品而上,何难? 吾欲利,而能谓人已耶? 北土既尔,他方复然,则后此数十年,松之布竟何所泄哉?(《农政全书》)

7)种复早,又偶值稀疏之处,偶遇肥饶之地,偶当丰稔之时,此四五事皆相得,则花王矣。然安能一一凑合若此,所谓万万中有一,而花王绝少也。(《农政全书》)

8)故曰:天地者,万物之上下也;阴阳者,血气之男女也;左右者,阴阳之道路也;水火者,阴阳之征兆也;阴阳者,万物之能始也。故曰:阴在内,阳之守也;阳在外,阴之使也。(《黄帝内经》)

9)天覆地载,万物方生。未出地者,命曰阴处,名曰阴中之阴;则出地者,命曰阴中之阳。阳予之正,阴为之主。故生因春,长因夏,收因秋,藏因冬,失常则天地四塞。阴阳之变,其在人者,亦数之可数。(《黄帝内经》)

10)错国于不倾之地者,授有德也。积于不涸之仓者,务五谷也。藏于不竭之府者,养桑麻育六畜也。下令如流水之原者,令顺民心也。(《管子》)

11)使民于不争之官者,使各为其所长也。明必死之路者,严刑罚也。开必得之门者,信庆赏也。(《管子》)

12)不为不可成者,量民力也。不求不可得者,不强民以其所恶也。不处不可久者,不偷取一时也。不行不可复者,不欺其民也。(《管子》)

2. 将以下两段话译成英文。

1)《天工开物》是中国古代百科全书式的科技著作,论述了农业和手工业两大领域内30个生产部门的技术,分上中下三卷,共18章,插图123幅。每章首有"宋子曰"一段作为引言,对全章内容作提要性叙述。上卷主要记述谷物栽培及农具、水利机械,养蚕与丝织技术,植物染料与染色技术,制盐技术与工具,甘蔗种植与制糖技术。中卷记述砖、瓦及白瓷的烧炼技术,冶炼与铸造技术,舟车结构与使用方法,锻造铁器的工艺,烧制石灰、采煤等技术,植物油脂提炼工艺,造纸技术。下卷记述金银等各种金属矿石的开采与冶炼技术,冷兵器的制造工艺,朱砂研制,制墨,酒曲制造,珍珠、宝石等的开采工艺等。

2)"社"是土地神,"稷"是五谷神。"社稷"是古代帝王、诸侯所祭祀的土地神和五谷神。土地神和五谷神是以农为本的汉民族最重要的原始崇拜物。古代君主为了祈求国事太平、五谷丰登,每年都要祭祀土地神和五谷神,"社稷"因此成为国家与政权的象征。

3. 以下一首诗选自《诗经》,是周王祭祖祀神的乐歌,根据释文进行翻译。

信彼南山,维禹甸之。畇畇原隰,曾孙田之。我疆我理,南东其亩。
上天同云,雨雪雰雰,益之以霡霂。既优既渥,既沾既足,生我百谷。
疆埸翼翼,黍稷彧彧。曾孙之穑,以为酒食。畀我尸宾,寿考万年!
中田有庐,疆埸有瓜。是剥是菹,献之皇祖。曾孙寿考,受天之祜。
祭以清酒,从以骍牡,享于祖考。执其鸾刀,以启其毛,取其血膋。
是烝是享,苾苾芬芬。祀事孔明,先祖是皇。报以介福。万寿无疆!

(《诗经·小雅·北山之什·信南山》)

【释文】

申展那终南山，只有禹来治理它。平整那高原和洼地，曾孙曾经种过它。我划疆界和治理，田亩从南从东我治它。

上天有阴云，下雪又纷纷。加上又小雨。既是水足又润渥，既经霑湿又满足，可以生长我百谷。

田地疆界很整饬，黍稷种得很密植。曾孙把它来收获，用作我们的酒食。给我作尸和宾客，神赐寿考万年值。

田中种得有萝卜，田边种得有杂瓜。是剥萝卜是腌瓜，献给皇祖不为差。曾孙因此得长寿，受天赐福得称嘉。

祭祀用的是清酒，跟着一头红牡牛，拿去献给先祖考。拿着他的鸾刀头，用来开脱它皮毛，取出它的血和油。

冬祭请神来受享，芬芬芳芳是馨香。祭祀的事很洁净，先祖受祭得安享。报祭用来赐大福，赐给万寿称无疆！

延伸阅读

《诗经》译文赏析

下文选自《诗经·周颂·闵予小子之什·载芟》。三个英译本分别为：威廉·詹宁斯（William Jennings）译*The Shi King*（以下简称“詹译”）；埃兹拉·庞德（Ezra Pound）译*The Classic Anthology Defined by Confucius*（以下简称“庞译”）；许渊冲译*Book of Poetry*（以下简称“许译”）。

载芟载柞，其耕泽泽。千耦其耘，徂隰徂畛。侯主侯伯，侯亚侯旅，侯彊侯以。有嗿其馌，思媚其妇，有依其士。有略其耜，俶载南亩。播厥百谷，实函斯活。驿驿其达，有厌其杰。厌厌其苗，绵绵其麃。载获济济，有实其积，万亿及秭。为酒为醴，烝畀祖妣，以洽百礼。有飶其香，邦家之光。有椒其馨，胡考之宁。匪且有且，匪今斯今，振古如兹。

詹译：

Clear the twitch-grass, clear the scrub;
Ploughs the soddened soil shall grub.
Thousand couples weed the ground,
Crossing swampy field and bound:
There the master, there the son,
Younger sons, aye every one;
Strong men here, assistants there.
Hear them o'er their (mid-day) fare.
Husbands eye their wives with pride,
Wives cling to their husbands' side.
Now the sharpened shares are in; —
On South Acres they begin.
Sown is grain, of every kind;
Living germs in all enshrined.

Bursting now, in faultless rows,
Succulent and tall it grows.

'Mid the young and thriving grain
Weeders wade, a numerous train.

Last, the reapers, band on band,
Pile the produce on the land,
Till the stacks unnumbered stand.

Liquor sweet and strong 'twill brew,
'Gainst the time when gifts be due
To departed dame and sire,
And for what all rites require.
Fragrant odour thence doth rise
That a nation glorifies;
While the pungent perfume chcers
Men in their declining years.

Not that here alone 'tis so,
Nor that now alone 'tis so: —
Thus it was long long ago.

庞译:

Gainst high scrub oak and rathe to plow the marsh
pair'd plowmen went, attacking gnarl and root;
pacing low slough and ordered boundry dyke
what crowd is here: the master and his son,
aids, wives and food; strong ready neighbour men
sharpened the plows, so came south fields a-grain.

A power from far and silent in the shoot
see how the spirit moves within the corn
strong as a stallion, quiet as water on tongue.
Here be the stalks a-row, silky and white;
wave as a cloth beneath the common sun.

Ordered the grain and rich beyond account,
fit to distil to drink in sacrifice.
Let manes come to taste what we devise
agnate and cognate. As pepper to ease old age
here without altar shall be holiness
not now for new, but as it was of old
tho' tongue be light against the power of grass.

许译：

The grass and bushes cleared away,
The ground is ploughed at break of day.
A thousand pairs weed, hoe in hand;
They toil in old or new-tilled land.
The master comes with all his sons,
The older and the younger ones.
They are all strong and stout;
At noon they take meals out.

They love their women fair
Who take of them good care.
With the sharp plough they wield,
They break the southern field.
All kinds of grain they sow
Burst into life and grow,
Young shoots without end rise;

The longest strike the eyes.
The grain grows lush here and there;
The toilers weed with care.
The reapers come around;
The grain's piled up aground.
There're millions of stacks fine
To be made food or wine
For our ancestors' shrine
And for the rites divine.
The delicious food
Is glory of kinghood.
The fragrant wine, behold!
Gives comfort to the old.
We reap not only here
But always in good cheer.
We reap not only for today
But always in our fathers' way.

拓展知识

1974年,《茶经》首个全译本 *The Classic of Tea* 出版,译者是时任美中贸易博物馆副馆长的弗朗西斯·罗斯·卡彭特。下文选自该译本引文部分,介绍了茶作为中国文化之魂的重要性。除脚注为本书编者所加外,其余均为原文摘录。

Tea: A Mirror of China's Soul

by Francis Ross Carpenter

To those of us used to thinking of today only as the raw material for tomorrow, Lu Yü's eighth-century *Classic of Tea* is a delicately refreshing suggestion of the importance of today itself. In reminding us of that, Lu Yü, his book and its subject reveal themselves as crystalline expressions of a uniquely Chinese attitude. Tea is, in fact, so integral to the Chinese spirit that it matters little how much is produced or drunk in other countries. It will always be China's drink.

In Lu Yü's China of the T'ang dynasty, the spirit of China and its body of belief were multifaceted, to be sure. But basic to that spirit was a conviction that every phenomenon of life, however small, was of compelling concern. Asked about the Tao—the supreme essence of spiritual reality—Chuang Tzu replied that there is no place that it is not. It is, he said, in the ant, the grass, the tiles and shards.

If the Highest Reality was a part of things so low, it followed that every act of living was an act of celebration in the festival of life. And the Chinese did celebrate life, intensely, passionately and continuously. One scholar has made the point that when Buddhism came to China, one reason that it found so receptive an audience was its doctrine of reincarnation. To the Indian Buddhist, reincarnation was an evil to be prevented. To the Chinese, on the other hand, an opportunity for a second engagement with life seemed a most appealing thought. Such an outlook was fundamental to a T'ang ethos in which every moment in time merited celebration as a part of the thread of life.

This is obviously not to say that the Chinese were an especially happy people. We need only refer to their poetry to appreciate their capacity for keen, deep and unremitting sadness. But surely great sorrow is no more than a reflection of one's capacity for great joy. It seems, at least, to be the case with a great body of Chinese poetry that it is either unabashedly sensual rejoicing in the moment or acute suffering over the evanescence of that moment. Consider "A Moonlight Night" by Liu Fang-p'ing, which attempts to freeze for eternity the coming of spring:

> The night has grown old and the moon only half gives us light.
> The Great Bear is at the horizon, the Smaller already set.
> This is the evening when I can believe in Spring
> As insect sounds filter through the warmth of my window silk of green.[①]

But by the same author and with many of the same images:

> Silk windows melt into twilight bronzes at sunset.
> In golden rooms bitter tears drop unseen
> As Spring skulks away from barren courtyards
> Where pear petals fall, blocking the doors.[②]

The joyous welcome to spring is balanced by the melancholy awareness of its impermanence. Every moment of life is life itself; every passing moment is the passing of a life. To the Chinese, it is the moment and the act of the moment that constitute life.

It was thus with tea, and so, said Lu Yü, the act of taking tea must be

---

① 此处引用的是唐代诗人刘方平的诗作《月夜》,原诗为:更深月色半人家,北斗阑干南斗斜。今夜偏知春气暖,虫声新透绿窗纱。

② 此处为刘方平的另一诗作《春怨》,原诗为:纱窗日落渐黄昏,金屋无人见泪痕。寂寞空庭春欲晚,梨花满地不开门。

attended by every device with which the Chinese celebrated life in its other forms. Every tea hour must become a masterpiece to serve as a distillation of all tea hours, as if it were the first and with no other to follow.

And so the act of drinking tea must be attended by beauty. Throughout the book, Lu Yü returns to that theme. The environment, the preparation, the ingredients, the tea itself, the tea bowl and the rest of the équipage must have an inner harmony expressed in the outward form. No matter that a certain porcelain be quite rare or quite expensive, if its color is wrong, it must be banished from the équipage. The tea must be chosen for its delicacy and the water for its purity. Even the equipment for manufacturing as well as that for brewing must reflect no lack of attention. Lu Yü is constantly cajoling his reader to "spruce it up" or to add some ornamentation "by way of decoration".

Another typical extension of the Chinese credo expressed by *The Classic of Tea* is the demanding ritual with which Lu Yü would surround the act of taking tea.

To the Chinese, particularly to the Confucianist, ritual was essential to the good life. It was not an end in itself, but it was—again—an outward form or a behavioral expression of an inward ethic. It was to some extent a reflection of the Chinese belief in the natural order of things. The "kowtow", a cause célèbre of so many western confrontations with the Chinese, was a ritualistic expression of such a belief. The younger brother kowtowed to the elder; the elder brother, to the father; the father, as subject, to the Emperor; and the Emperor, as Son of Heaven, to Heaven itself. There was nothing degrading in the act. It was simply ritualistic reassurance that the natural order was still intact.

As ritual was the behavioral expression of an ethic, so did it help to intensify one's belief in the ethic. Ritualistic deference paved the way for genuine respect. Ritualistic acts of graciousness or politeness showed the path to peace and harmony and love. In many ways, ritual served the same ultimate purposes as law in the West.

To the Chinese, especially the Confucianists, ritual was a step toward freedom and not a retreat. Self-discipline, they would have told us, is the first step toward self-realization. And the imposition of outward forms is the first step toward self-discipline. The absence of form and ritual and rules does not assure freedom to the undisciplined mind. Their presence does not deny freedom, but rather abets its attainment, for the disciplined one.

Neither was it intended as a sterile imitation of past forms but as a means of finding what past sages had found or at least searched for. Ritual provided the order prerequisite to that search.

By the same token, ritual provided a context in which beauty was possible. The order in the équipage, in the preparation when it came time for tea, in the appointments ritualistically achieved, invited harmony of host with guest and guest with quest. And where harmony is, the Chinese believed, beauty will reign. It is, nonetheless, important to emphasize that ritual practice assured none of those ends. It merely created a context in which achievement of the ends (of beauty, or truth, or self-realization) was possible.

Inevitably, then, Lu Yü insists throughout his work on the importance of ritual when one drinks tea. For him the ritual is decisively important. It is another way of celebrating another act of living. The water must be boiled so that it goes through the appropriate stages. The tea must be tested and tasted before it is selected for steeping. There are nine stages through which tea must pass during manufacture and seven during brewing. There are twenty-four implements and each must be used each time (with a set of closely prescribed exceptions). So important is the ritual that it is better, Lu says, to dispense with the tea if even one of the twenty-four implements is missing. When the tea is plucked, where the water is chosen, who is invited to share the tea are of enormous importance. Lu Yü considered it almost unthinkable that a guest should fail to appear. Should a guest be missing, he warns, the quality of the tea must be such as to atone.

One can find yet another of the traditional parts of the credo in the whole question of moderation. It is to be found both in Lu's advocacy of moderation and in his praise of tea for the moderation it induces. Always sip tea, he advises, as if tea were life itself, and do not dissipate the flavor. Never take more than three cups unless you are quite thirsty, for tea, he says, is the very essence of moderation and helps to still the six passions. In a real sense then, moderation is simply another dimension of the love of order in which the passions are restrained, the "middle way" achieved and the beauty of the resultant harmony revealed. That moderation is more easily attained through gentle tea is one of the reasons that it will always be a Chinese drink whatever its origin.

Finally, and perhaps most important, tea epitomizes the Chinese attitude toward time and change. To the Westerner, life has always seemed very linear. Today must somehow improve upon yesterday and tomorrow must extend and build upon today. The Chinese have taken a more cyclic view of their world. The Westerner has danced to an insistent beat called "progress," the Chinese, to a rhythm of natural movement.

Natural movement had several implications for the early Chinese, none of which denied the fact of change. One such implication was that standstill and rest are a part of change. The growth and development of what is unnatural (and therefore wrong) is not change but its very antithesis. Central to the Chinese concept of change is the idea of return. To get back to the starting point and the source of one's strength was of the essence of change. "Fill a bowl to the brim and you will regret it / A sword too sharply ground will dull,"[①] says the *Tao Tê Ching* in a warning not to take progress beyond its natural limits. Even death constituted a return and was a part of the natural order. It was neither to be feared nor opposed but rather to be accepted as a fact of life out of which new life would evolve. When Tzu-lai lay dying, his wife and children were admonished to hush their wailing. "Do not disturb him in his change," said his friend Tzu-li[②].

Implicit in *The Classic of Tea* is the acceptance of this idea of change. That there must be a time to rest is a lesson taught by nature. We disregard it at society's peril, and Lu Yü understood that. Hence, tea was not just a medicine to banish drowsiness (although he frequently praises it for that virtue). It was a means of helping man return to his starting point—that hour in the rhythm of the day when the prince and the peasant shared thoughts and a common cheer as they readied themselves for their separate lives.

Such are the virtues that Lu Yü finds in tea and it is those virtues that dictate the loving attention he would have his reader give to the moment of tea.

Not everyone in China accepted all Lu Yü's ideals. The Confucian, for example, with a prescribed ritual for mourning the dead, would have been shocked at Chuang Tzu who refused to mourn his wife's death, saying only that life had evolved death and he would not impose upon her rest. China and the

① 原文为:持而盈之,不如其已。揣而锐之,不可长保。

② 子来和子犁为《庄子》中的故事人物。

Chinese were compounded of many parts, philosophically, socially, economically. There were the Confucian official, the Taoist adept and the Buddhist priest; the scholar and peasant, the artisan and merchant; the rich landlord and humble farmer.

Lu Yü did not ignore the differences. Rather he accepted and transcended them. In his treatise, the author, the book, and its subject, which is tea, become more than the sum of all those parts. They are, in a word, Chinese.

练　习

1. What is the main theme of the passage?

   A. The history of tea in China.

   B. The importance of ritual in Chinese culture.

   C. The tea spirit in Chinese culture.

   D. The different types of tea in Chinese culture.

2. According to the passage, what was the attitude of the Chinese towards life?

   A. Life should be lived in pursuit of progress.

   B. Life should be celebrated in every moment.

   C. Life should be focused on material possessions.

   D. Life should be avoided due to its impermanence.

3. What is the significance of ritual in Chinese culture, as mentioned in the passage?

   A. Ritual is a way to control people's behavior.

   B. Ritual is essential for self-discipline and self-realization.

   C. Ritual is a form of entertainment in Chinese society.

   D. Ritual is a way to show off one's wealth and status.

4. According to Lu Yü, why is beauty important in the act of drinking tea?

A. Beauty reflects the inner harmony of the tea ceremony.

B. Beauty enhances the flavor of the tea.

C. Beauty impresses guests and shows off wealth.

D. Beauty is a distraction from the act of drinking tea.

5. How does tea embody the Chinese attitude towards time and change?

A. Tea represents progress and improvement.

B. Tea signifies the inevitability of death.

C. Tea symbolizes the cyclical nature of life.

D. Tea promotes constant change and evolution.

6. What does Lu Yü emphasize about moderation in the act of drinking tea?

A. Moderation is essential for preventing intoxication.

B. Moderation helps to enhance the flavor of the tea.

C. Moderation is a key aspect of Chinese culture.

D. Moderation is necessary for self-discipline and harmony.

7. How does Lu Yü view the concept of change in Chinese society?

A. Change should be resisted and avoided at all costs.

B. Change should be embraced as a natural part of life.

C. Change should be controlled through strict rituals.

D. Change should be feared and opposed at all times.

8. How does Lu Yü transcend the differences in Chinese society in his treatise on tea?

A. By ignoring the differences and focusing solely on tea.

B. By imposing his own beliefs and ideals on all Chinese people.

C. By accepting and incorporating the diverse elements of Chinese culture.

D. By excluding certain groups of people from participating in tea ceremonies.

9. Which of the following statements is NOT true according to the passage?

A. The equipment for manufacturing and brewing tea is not of importance.

B. Not everyone in China accepted all Lu Yü's ideals.

C. To the Chinese, ritual was a step toward freedom and not a retreat.

D. Moderation is simply another dimension of the love of order.

10. What is the overall message conveyed by the passage?

A. Tea is the most important aspect of Chinese culture.

B. Chinese society is rigid and unchanging in its traditions.

C. The Chinese attitude towards life is focused on progress and improvement.

D. The act of drinking tea reflects the essence of Chinese philosophy.

参考文献

[1] Birrell, Anne. trans. The Classic of Mountains and Seas 山海经[M]. London: Penguin Books, 1999.

[2] Bloom, Irene. trans. Mencius 孟子[M]. New York: Columbia University Press, 2011.

[3] Carpenter, Francis Ross. trans. The Classic of Tea 茶经[M]. Boston: Little, Brown and Company, 1974.

[4] Jennings, William. trans. The Shi King 诗经[M]. London: George Routledge and Sons Ltd, 1891.

[5] Knoblock, John, Jeffrey Riegel. trans. The Annals of Lü Buwei 吕氏春秋[M]. Stanford: Stanford University Press, 2000.

[6] Legge, James. The Chinese Classics: With a Translation, Critical and Exegetical Notes, Prolegomena, and Copious Indexes[M]. Hong Kong: The Author's, 1871.

[7] Major, John S., et al. trans. The Huainazi 淮南子[M]. New York: Columbia University Press, 2010.

[8] Munday J. Introducing Translation Studies [M]. London and New York: Routledge, 2016:124.

[9] Needham, Joseph. Science and Civilization in China [M]. London: Cambridge University Press, 1984.

[10] Nida, Eugene. Linguistics and Ethnology in Translation Problems [J]. Word,

1945, 1(2): 194–208.

[11] On Fan Sheng-chih Shu 氾胜之书今释[M]. 石声汉，译. 北京：科学出版社，1959.

[12] Pound, Ezra. trans. The Classic Anthology Defined by Confucius 诗经[M]. London: Faber and Faber Ltd, 1974.

[13] Shang Shu 尚书(英汉对照)[M]. 周秉钧，译. 长沙：湖南人民出版社，2013：5.

[14] Shaw C. Directions for the Cultivation of Cotton[J]. The Chinese Repository, 1849, 18(9): 449–469.

[15] Shi S. H. A Preliminary Survey of the Book: Chi Min Yao Shu[M]. Beijing: Science Press, 1962.

[16] Sung Y. T'ien-kung K'ai-wu: Chinese Technology in the Seventeenth Century [M]. E-u Zen Sun and Shiou-Chuan Sun. Trans. London: The Pennsylvania State University Press, 1966.

[17] The Book of Rites 礼记[M]. Legge, James, 译. 郑州: 中州古籍出版社, 2016.

[18] The Classic of Mountains and Seas 山海经[M]. 陈成，王宏，赵峥，译. 长沙：湖南人民出版社，2010.

[19] The Classic on Tea. 茶经[M]. 邱贵溪，译. 上海：上海交通大学出版社，2023.

[20] W-Allyn Rickett. GUANZI—Political, Economic, and Philosophical Essays from Early China [M]. Princeton: Princeton University Press, 1985.

[21] William Jennings. The Shi King, the Old Poetry Classic of the Chinese [M]. General Books LLC, 2010.

[22] Xunzi Vol. Ⅰ, Ⅱ 荀子(全二册)[M]. 张觉, 译. 长沙: 湖南人民出版社, 1999.

[23] Yellow Emperor's Canon of Medicine: Plain Conversation. Vol. Ⅰ, Ⅱ, Ⅲ. 黄帝内经·素问(全三册)[M]. 李照国，刘希茹，译. 西安：世界图书出版公司，2005.

[24] 陈旉.陈旉农书校释[M].刘铭,校释.北京:中国农业出版社,2015.
[25] 纯道.禅艺茶道[M].上海:文汇出版社,2017:78-79.
[26] 戴圣.礼记[M].James Legge,译.郑州:中州古籍出版社,2016:7.
[27] 丁亮,李善兰.《植物学》译介及其影响研究[D].南京:南京信息工程大学,2015.
[28] 都贺庭钟.天工开物菅生堂本序.菅本书首[M].大阪:菅生堂,1771.
[29] 樊志民.陕西古代农田水利科学技术初探[J].水资源与水工程学报,1990(3):84-91.
[30] 范文来.《齐民要术》中的中国古代酿酒技术[J].酿酒,2020,47(6):111-113.
[31] 方梦之.译学辞典[W].上海:上海外语教育出版社,2004.
[32] 方梦之,庄智象.中国翻译家研究[M].上海:上海教育出版社,2017.
[33] 费振玠,曹洸.从《天工开物》外译情况谈科技翻译[J].上海科技翻译,1988(2):41-43.
[34] 付雷.晚清中下学生物教科书出版机构举隅[J].科普研究,2014(6):61-72.
[35] 傅金泉.中国古代科学家对酿酒发酵化学的重大贡献[J].酿酒,2012,39(5):87-91.
[36] 管仲.Guanzi Vol. Ⅰ, Ⅱ, Ⅲ, Ⅳ 管子(全四册)[M].翟江月,译.桂林:广西师范大学出版社, 2005.
[37] 郭超,夏于全.传世名著百部[M].北京:蓝天出版社,1998.
[38] 韩忠治.《农政全书》词汇研究[D].河北师范大学,2015.
[39] 韩忠治.《农政全书》中的朴素农业生态观[N].学习时报,2022-06-28.
[40] 何善芬.英汉语言对比研究[M].上海:上海外语教育出版社,2002:466-477.
[41] 胡火金.日本学者对中国农业史的研究[J].史学月刊,2016(11):124-129.
[42] 华北农业大学农业科学技术史研究组.精耕细作——我国古代农业科学技术的优良传统(一)[J].中国农业科学,1978(1):91-96.
[43] 贾思勰. A Preliminary Survey of the Book Ch'i Min Yao Shu 2nd ed. 齐民要术

概论(第二版)[M].石声汉,译.北京:科学出版社,1962.

[44] 贾思勰.齐民要术(全二册)[M].石声汉,校.北京:中华书局,2015.

[45] 贾思勰.齐民要术今释[M].石声汉,译.北京:中华书局,2009.

[46] 姜望琪.论术语翻译的标准[J].上海翻译,2005(S1):80-84.

[47] 孔令翠,刘芹利.中国农学典籍译介梳理与简析[J].当代外语研究,2019(4):106-114.

[48] 孔令翠.农学典籍《氾胜之书》的辑佚、今译与自译“三位一体”模式研究[J].外语与翻译,2019,26(3):47-52.

[49] 李海军.18世纪以来《农政全书》在英语世界译介与传播简论[J].燕山大学学报(哲学社会科学版),2017,18(6):33-37,43.

[50] 李时珍. Condensed Compendium of Materia Medica. Vol. Ⅰ, Ⅱ, Ⅲ, Ⅳ, Ⅴ. 本草纲目选(全五册)[M].罗希文,译.北京:外文出版社,2012.

[51] 李宇明.谈术语本土化、规范化与国际化[J].中国科技术语,2007(4):5-10.

[52] 林煌天,贺崇寅.中国科技翻译家辞典[M].上海:上海翻译出版公司,1991.

[53] 刘安. Huai Nan Zi Vol. Ⅰ, Ⅱ, Ⅲ. 淮南子(全三册)[M].翟江月,牟爱鹏,译.桂林:广西师范大学出版社,2010.

[54] 刘瑾.文化强国背景下农业典籍中的哲学智慧对外传播路径研究[J].中国油脂,2023(10).

[55] 刘欣.探析古代酿酒技术[J].黑龙江史志,2015(5):295.

[56] 刘性峰,王宏.中国古典科技翻译研究架构建[J].上海翻译,2016(4):72,77-81,94.

[57] 刘学礼.西方生物学的传入与中国近代生物学的萌芽[J].自然辩证法通讯,1991(6):43-52.

[58] 陆羽,陆廷灿. The Classic of Tea. The Sequel to The Classic of Tea Vol. Ⅰ, Ⅱ. 茶经.续茶经(全二册)[M].姜怡,姜欣,译.长沙:湖南人民出版社,2009.

[59] 陆羽,陆廷灿.茶经.续茶经(全二册)[M].杜斌,译.北京:中华书局,2020.

[60] 罗桂环.我国早期的两本植物学译著——《植物学》和《植物图说》及其术语[J].自然科学史研究,1987(4):383-387.

[61] 罗选民,李婕.典籍翻译的内涵研究[J].外语教学,2020,41(6):83-88.

[62] 吕不韦. The Spring and Autumn of Lü Buwei Vol. Ⅰ,Ⅱ,Ⅲ. 吕氏春秋(全三册)[M].翟江月,译.桂林:广西师范大学出版社,2005.

[63] 吕不韦.Lü's Commentaries of History 吕氏春秋[M].汤博文,译.北京:外文出版社,2010.

[64] 马清海.试论科技翻译的标准和科技术语的翻译原则[J].中国翻译,1997(1):2.

[65] 马勇.青梅煮酒论英雄:马勇评近代史人物[M].南昌:江西人民出版社,2014.

[66] 梅阳春.古代科技典籍英译——文本、文体与翻译方法的选择[J].上海翻译,2014(3):70-74.

[67] 梅阳春.西方读者期待视阈下的中国科技典籍翻译文本构建策略[J].西安外国语大学学报,2018(3):102-106.

[68] 农政全书校注(全三册)[M].石声汉,校.北京:中华书局,2020.

[69] 潘才宝,刘湘溶.《齐民要术》的生态伦理取向探究[J].长春理工大学学报(社会科学版),2018,31(4):57-60.

[70] 潘吉星.《天工开物》在国外的传播和影响[N].北京日报,2013-01-29.

[71] 潘吉星.宋应星评传[M].南京:南京大学出版社,1990.

[72] 潘吉星.徐光启著《农政全书》在国外的传播[J].情报学刊,1984(3):94-96.

[73] 沈括.梦溪笔谈全译[M].王宏,赵峥,胡道静,译.上海:上海古籍出版社,2013.

[74] 沈括.梦溪笔谈[M].诸雨辰,译注.北京:中华书局,2016.

[75] 施由明.论中国茶叶向世界传播对世界文明的贡献[J].农业考古,2018(5):7-12.

[76] 石声汉.中国古代农书评介[M].北京:农业出版社,1980.

[77] 宋应星. Tian Gong Kai Wu 天工开物[M].王义静,王海燕,刘迎春,等译.广州:广东教育出版社,2011.

[78] 宋应星.天工开物[M].杨维增,译注.北京:中华书局,2021.

[79] 孙美娟.中国社会科学网-中国社会科学报[N].2022-06-22.

[80] 孙星衍.神农本草经(汉英对照)[M].刘希茹,李照国,译.上海:上海三联书店,2017.

[81] 孙雁冰.晚清(1840—1912)来华传教士植物学译著及其植物学术语研究[J].山东科技大学学报(社会科学版),2019,21(6):33-38.

[82] 谈敏.法国重农学派学说的中国渊源[M].上海:上海人民出版社,2010.

[83] 汪榕培,王宏.中国典籍翻译[M].上海:上海外语教育出版社,2009.

[84] 汪子春.我国传播近代植物学知识的第一部译著《植物学》[J].自然科学史研究,1984(1):90-96.

[85] 王翠.论新时代中国农学典籍的翻译与传播[J].南京工程学院学报(社会科学版),2019,19(4):18-23.

[86] 王宏,刘性峰.国内近十年《茶经》英译研究(2008-2017)[J].外文研究,2018,6(2):64-69,108.

[87] 王惠琼,孔令翠.20世纪前海外中国农业科技典籍译介研究[J].外国语文,2022,38(3):108-115.

[88] 王思明,李昕升.农业文明:丝绸之路上“行走”的种子[N].中国社会科学报,2017-03-02.

[89] 王小凤,张沉香.科技英语翻译过程的多维思索[J].中国科技翻译,2006,(4):33-36.

[90] 王秀丽.《茶经》语言的审美特质[J].农业考古,2018(2):170-172.

[91] 王烟朦.基于《天工开物》的中国古代文化类科技术语英译方法探究[J].中国翻译,2022,43(2):156-163.

[92] 王燕,李正栓.《大中华文库》科技典籍英译与中国文化对外传播[J].上海翻译,2020(5):53-57,94.

[93] 熊兵.翻译研究中的概念混淆—以"翻译策略""翻译方法"和"翻译技巧"为例[J].中国翻译,2014,35(3):82-88.

[94] 徐光启.农政全书[M].上海:上海古籍出版社,2010.

[95] 徐玉凤.知识翻译学视域下《齐民要术》的数字英译[J].当代外语研究,2023,468(6):46-53.

[96] 许明武,王烟朦.中国科技典籍英译研究(1997-2016):成绩、问题与建议[J].中国外语,2017,14(2):96-103.

[97] 闫畅,王银泉.中国农业典籍英译研究:现状、问题与对策(2009-2018)[J].燕山大学学报(哲学社会科学版),2019,20(3):49-58.

[98] 杨勇.试论《齐民要术》中的我国古代制曲、酿酒发酵技术[J].西北农林科技大学学报(自然科学版),1985(4):55-64.

[99] 杨直民.农业科学技术史研究的蓬勃发展[J].中国农史,1985(3):55-63.

[100] 佚名.Book of Poetry 诗经[M].许渊冲,译.北京:五洲传播出版社,2011.

[101] 袁慧,冯炜.基于目的顺应论的农学典籍《齐民要术》科技术语翻译研究[J].长春理工大学学报(社会科学版),2023,36(6):158-162.

[102] 袁梦瑶,董晓波.《茶经》译介推动中国茶文化走向世界[N].中国社会科学报,2019-03-15.

[103] 张保国,周鹤.石声汉的农学典籍译介模式及其启示[J].解放军外国语学院学报,2022,45(5):119-127.

[104] 张保国.中国农学典籍《齐民要术》译介研究述评[J].乐山师范学院学报,2020,35(3):46-51.

[105] 张帆.安徽大农业史述要[M].合肥:中国科学技术大学出版社,2011:106.

[106] 张芳,王思明,等.中国农业古籍目录[M].北京:国家图书馆出版社,2003.

[107] 张翮.基于"双语对校"的晚清译著《植物学》研究[D].合肥:中国科学技术

大学,2017.

[108] 张萌,刘俊仙.丝绸之路与古代中国蚕桑技术的外传[J].中国民族博览,2015(8):184-186.

[109] 张胜忠.古代醋名小识[J].吉林中医药,1992(2):46-47.

[110] 张志聪.本草崇原(汉英对照)[M].孙慧,译.苏州:苏州大学出版社,2021.

[111] 赵丽梅,汪剑.认知翻译学视角下《神农本草经》的英译研究[J].中国中医基础医学杂志,2022,28(8):1335-1338.

[112] 周魁一.中国古代农田灌溉排水技术[J].古今农业,1997,121(1):1-12.

[113] 周倩儒.论《植物学》术语译名的民族化[J].新楚文化,2022(10):61-64.

[114] 周有光.漫谈科技术语的民族化与国际化[J].中国科技术语,2010(1):8-10.

[115] 朱高正.中国文化对西方的影响[J].自然辩证法研究,2001(8):49-55.

[116] 朱肱,顾宏义,任仁仁.北山酒经(外十种)[M].上海:上海书店出版社,2016.

[117] 朱橚.《救荒本草》汉英对照[M].范延妮,译.苏州:苏州大学出版社,2019.